Pierre Auger, Max Born,
Werner Heisenberg, Erwin Schrödinger

ON FUNDAMENTAL PHYSICS
AND
SCIENTIFIC KNOWLEDGE

Edited by Vesselin Petkov

MINKOWSKI
Institute Press

Pierre Auger (May 14, 1899 – December 25, 1993)
Max Born (11 December 1882 – 5 January 1970)
Werner Heisenberg (5 December 1901 – 1 February 1976)
Erwin Schrödinger (12 August 1887 – 4 January 1961)

Cover photos:
https://www.biografiasyvidas.com/biografia/a/aug
er.htm;
https://en.wikipedia.org/wiki/Max_Born#/media/Fi
le:Max_Born.jpg
https://en.wikipedia.org/wiki/Werner_Heisenber
g#/media/File:Bundesarchiv_Bild183-R57262,_Werne
r_Heisenberg.jpg;
https://www.nobelprize.org/prizes/physics/1933/s
chrodinger/biographical/

Published 2021

ISBN: 978-1-989970-35-5 (softcover)
ISBN: 978-1-989970-36-2 (ebook)

Minkowski Institute Press
Montreal, Quebec, Canada
http://minkowskiinstitute.org/mip/

For information on all Minkowski Institute Press
publications visit our website at
http://minkowskiinstitute.org/mip/books/

EDITOR'S PREFACE

This volume contains new publications of

- the book: Heisenberg, Born, Schrödinger and Auger, *On Modern Physics**

- Werner Heisenberg, The development of quantum mechanics, Nobel Lecture, December 11, 1933

- Erwin Schrödinger, The fundamental idea of wave mechanics, Nobel Lecture, December 12, 1933

- Max Born, The statistical interpretation of quantum mechanics, Nobel Lecture, December 11, 1954

In this collection three Nobel laureates and a renowned authority on space exploration discuss a wide range of issues – from lessons that can be learned from the ancient Greek philosophers, to the advancements in fundamental physics in the twentieth century, to the dark implications of scientific discoveries, to the methods and limits of scientific knowledge – in a language that is understandable by a wide audience. The collections ends with the Nobel lectures of Heisenberg, Schrödinger and Born.

*Heisenberg, Born, Schrödinger and Auger, *On Modern Physics* (Collier Books, New York 1962) is the English translation by M. Goodman and J. W. Binns of the original Italian publication Heisenberg, Born, Schrödinger, Auger, *Discussione sulla fisica moderna* (Paolo Boringhieri Editore, Turin 1960).

The texts were typeset in LaTeX by Svetla Petkova and noticed typos were corrected. A number of editor's notes were made in the text of the book *On Modern Physics*.

2 December 2021

Vesselin Petkov
Minkowski Institute
Montreal

Contents

On Modern Physics*

*New publication of Heisenberg, Born, Schrödinger and Auger, *On Modern Physics* (Collier Books, New York 1962), which is the English translation by M. Goodman and J. W. Binns of the original Italian publication Heisenberg, Born, Schrödinger, Auger, *Discussione sulla fisica moderna* (Paolo Boringhieri Editore, Turin 1960).

1

Planck's Discovery and the Philosophical Problems of Atomic Physics[1]

Werner Heisenberg

At a world conference on atomic energy, such as the one now taking place here in Geneva, where we consider the enormous amount of work devoted to the development of atomic physics in the most diverse countries, where we hear of the hundreds of projects attempting to apply the theoretical results of atomic physics to industrial purposes, we run the risk of overlooking, in the mass of details, the fact that our purpose today is to solve problems that have faced humanity for a very long time and that the theoretical work of our era is related to the efforts undertaken by man thousands of years ago. In today's lecture we shall speak of these broad historical relationships. At first glance they are undoubtedly more interesting to the historian than to the physicist, but on reflection the physicist, too, can observe certain governing principles that allow him a valuable insight into his own present problems.

Modern physics and particularly the quantum theory discovered by Planck, the centenary of whose birth is celebrated this year, have raised a series of very general questions, which not only concern the narrower problems of

---

[1]Lecture delivered September 4, 1958.

physics but also deal with the method of the exact natural sciences and with the nature of matter. These questions have compelled the physicist to reconsider the philosophical problems that seemed to have been definitely answered in the narrow framework of classical physics.

There are two groups of problems in particular that have again been raised by Planck's discovery, and these will be the subject of today's lecture.

One concerns the essence of matter or, more exactly, the Greek philosophers' old question of how it is possible to reduce to simple principles the motley and manifold phenomena surrounding matter and thus make them intelligible. The other concerns an epistemological problem, which particularly since Kant has been repeatedly raised, of how far it is possible to objectify our observations of nature – or our sensory experience in general – that is, to determine from observed phenomena an objective process independent of the observer. Kant had spoken of "the thing in itself." He was later often accused, even from the philosophical viewpoint, of inconsistency in this concept of "the thing in itself" in his philosophy. In the quantum theory this problem of the objective background of phenomena has arisen in a new and very surprising form. This question can therefore also be treated anew from the standpoint of the modern natural sciences.

I. In today's lecture we shall first deal with the problems in the realm of natural philosophy resulting from the search for a unified understanding of material phenomena. The Greek philosophers, meditating on the basis of visible phenomena, came up against the question of the smallest particles of matter. The result of this search was that, at the end of that period of human thought, there were two opposing concepts that have exercised the strongest influence on later developments of philosophy. These concepts were labeled "materialism" and "idealism."

The atomic theory founded by Leucippus and Democritus considered the smallest particles of matter as "that which exists" in the strictest sense. These particles were considered as indivisible and unalterable. They were eternal and ultimate units; they were therefore called atoms, and they neither needed nor had any further explanation. They had no properties other than geometrical ones. According to the philosophers, they did have a definite form. They were separated from each other by empty space and, by their different positions or movements in empty space, could produce a large variety of phenomena, but they had neither color nor smell or taste, still less temperature or other physical properties familiar to us. The properties of things that we perceive were indirectly caused by the different arrangement and movement of the atoms. Just as tragedy and comedy can be written with the same letters, so the most diverse happenings in the world can, according to the doctrine of Democritus, be realized by the same atoms. The atoms were therefore the true, objectively real core of matter and thus of all phenomena. They were, as I have already said, "that which exists" in the strictest sense, while the large variety of phenomena were produced only indirectly by the atoms. Therefore, this concept was called materialism.

For Plato, on the other hand, the smallest particles of matter are, so to speak, only geometrical forms. Plato considers the smallest particles of the elements identical with the regular bodies of geometry. Like Empedocles, he assumes the four elements to be earth, water, air, and fire. He conceives the smallest particles of the elements earth as cubes, the smallest particles of the element water as icosahedrons; similarly, he imagined the elementary particles of fire as tetrahedrons and those of air as octahedrons. The form is characteristic for the properties of the element. In contrast to Democritus, the smallest particles are not unalterable or indestructible for Plato; on the contrary they can be resolved into triangles and can be reconstructed from triangles. In this theory they are therefore not called atoms. The triangles themselves are no longer matter, for they have no spatial dimensions. Therefore, in Plato, at the lowest limit of the series of material structures, there is really no longer anything material, but a mathematical form if you like, an intellectual construct. The ultimate root from which the world can be uniformly understood is, in Plato, mathematical symmetry, the image, the idea; this concept is therefore called idealism.

It is remarkable that this old question of materialism and idealism has again been raised in a very definite form by modem atomic physics and particularly by the quantum theory. Before the discovery of Planck's action quantum the modem exact natural sciences, physics and chemistry, were materialistically oriented. In the nineteenth century the atoms of chemistry and their constituents, which to-day are called elementary particles, were considered as the only thing that really exists, as the real substratum of all matter. The existence of atoms neither had nor needed further explanation.

Planck, however, had discovered a quality of discontinuity in radiation phenomena, which seemed related in a surprising manner to the existence of atoms, but could on the other hand not be explained on the basis of their existence.

This characteristic, revealed by the action quantum, led to the idea that discontinuity as well as the existence of the atom could be joint manifestations of a fundamental law of nature, of a mathematical structure in nature, and that its formulation might lead to a unified understanding of the structure of matter, which the Greek philosophers had sought. The existence of atoms was therefore perhaps not an ultimate fact, incapable of further explanation. This existence could be ascribed, as in Plato, to the action of mathematically formulable laws of nature, that is, to the effect of mathematical symmetries.

Planck's law of radiation also differs in a very characteristic manner from the previously formulated laws of nature. Although the previous laws of nature, e.g., Newtonian mechanics, contained so-called constants, these constants referred to the properties of objects, e.g., their mass or the intensity of the force acting between two bodies or the like. On the other hand, Planck's action quantum, which is the characteristic constant in his law of radiation, does not represent a property of objects, but a property of nature. It establishes a scale in nature and demonstrates at the same time that, under conditions in which the effects are very large in comparison with Planck's action quantum (as in all phenomena of daily life), natural phenomena take a different course than in those cases in which the effects are of the order of atomic size, that is, of Planck's quantum. While the laws of former physics, e.g., Newtonian mechanics, should basically be equally valid for all orders of magnitude (the movement of the moon around the earth should obey the same laws as the fall of the apple from the tree or the deviation of an alpha particle that grazes the nucleus of an atom), Planck's law of radiation shows for the first time that there are scales in nature, that phenomena in different ranges of magnitude are not necessarily of the same type.

Already a few years after Planck's discovery, the significance of a second "measurement constant" was understood. Einstein's special theory of relatively made it clear to physicists that the velocity of light did not, as had previously been supposed in electrodynamics, describe the property of a special substance – "ether" – that supported the propagation of light, but that a property of space and time was involved, that is, a general property of nature not related in any way to particular objects or things in nature. Thus, the velocity of light can also be considered as a measurement constant of nature.

Our intuitive concepts of space and time can be applied

only to those phenomena in which small velocities with respect to the velocities of light are involved. Conversely, the well-known paradoxes of the theory of relativity are based on the fact that phenomena involving velocities near that of light cannot be properly interpreted with our normal concept of space and time. May I remind you of the well-known paradox of the clocks – that for a rapidly moving observer time apparently moves more slowly than for a stationary one. After the mathematical structure of the special theory of relativity had been made clear, it very soon became possible in the first decade of this century to analyze the physical significance of these mathematical relationships. This was done so thoroughly that it was possible to understand completely the aspects of nature connected with the velocity of light as a measurement constant. The many discussions on the theory of relativity clearly show that our deep-rooted concepts impeded the understanding of the theory, but the objections were rapidly overcome.

II. It was, however, much more difficult to understand the physical relationships connected with the existence of Planck's action quantum. It appeared probable from a paper of Einstein in the year 1918 that the laws of the quantum theory in some way or other involved statistical relationships. But the first attempt to thoroughly study the statistical nature of the laws of the quantum theory was made by Bohr, Kramers and Slater in 1924. The relationship between electromagnetic fields, which had been considered to be the propagators of light in classical physics since Maxwell, and the discontinuous, i.e., quantumwise, absorption and emission of atoms as postulated by Planck was interpreted in the following manner: The field of electromagnetic waves, to which the phenomena of interference and diffraction are manifestly due, determines only the probability that an atom will absorb or emit light energy by quanta in the space under consideration. The magnetic field was thus no longer considered as a field of force that acts on the electric charge of the atom and causes movement. Its action takes place more indirectly: the field determines only the probability that emission or absorption takes place.

Later this interpretation was shown to be not quite exact. The actual relationships were still somewhat uncertain; somewhat later they were correctly formulated by Born. Nevertheless, the work of Bohr, Kramers and Slater contained the decisive concept, that the laws of nature determine not the occurrence of an event, but the probability that an event will take place, and that the probability must be related to a wave field that obeys a mathematically formulable wave equation.

This was a decisive step away from classical physics; basically a concept that played an important part in Aristotle's philosophy was used. The probability waves of Bohr, Kramers and Slater can be interpreted as a quantitative formulation of the concept of δύναμις, possibility, or in the later Latin version, 'potentia,' in Aristotle's philosophy. The concept that events are not determined in a peremptory manner, but that the possibility or 'tendency' for an event to take place has a kind of reality – a certain intermediate layer of reality, halfway between the massive reality of matter and the intellectual reality of the idea or the image – this concept plays a decisive role in Aristotle's philosophy. In modem quantum theory this concept takes on a new form; it is formulated quantitatively as probability and subjected to mathematically expressible laws of nature. The laws of nature formulated in mathematical terms no longer determine the phenomena themselves, but the possibility of happening, the probability that something will happen.

This introduction of probability corresponded at first quite closely to the situation found in experiments with atomic phenomena. If the physicist determines the intensity of a radioactive radiation by counting how often this radiation activates the tube in a given time, he admits implicitly that the intensity of radioactive radiation regulates the probability of the counter's responding. The exact interval of time between impulses does not interest the physicist – he says they are statistically distributed. What matters is only the average frequency of the impulses.

The fact that this statistical interpretation reproduces exactly the experimental situation has been proved in many investigations. Quantum mechanics has also obtained exact confirmation in experiments that permit quantitative evidence, as, for example, on the wave length of spectrum lines or the binding energy of molecules. There could be no doubt of the correctness of the theory.

The problem of the compatibility of this statistical interpretation with the large store of experience collected in so-called classical physics was, however, more difficult. All experiments depend on an unequivocal relation between the observation and the physical phenomena on which it is based. If, for example, we measure a spectrum line of a definite frequency with a diffraction grating, we take it for granted that the atoms of the radiating substance must have emitted light of that frequency. Or, if a photographic plate is blackened, we suppose that it has been struck at that point by rays or particles of matter. Physics, in collecting experimental data, thus utilizes the unequivocal determinateness of events and thus apparently finds itself somewhat opposed to the experimental situation in the atomic field and to the quantum theory. It is precisely here that this unequivocal determinateness of events is questioned.

This apparent inner contradiction is eliminated in modern physics by establishing that the determinateness of phenomena exists only insofar as they are described with the concepts of classical physics. The application of these concepts is on the other hand limited by the so-called uncertainty relationships; these contain quantitative data on the limits that are placed on the application of classical concepts. The physicist thus knows in which cases he may consider events as determined and in which cases he may not; he can consequently use a method devoid of intrinsic contradictions for the observation and its physical interpretation. Of course, the question arises why it is still necessary to use the concepts of classical physics, why it is not possible to transform the whole physical description to a new system of concepts based on the quantum theory.

Here it is first of all necessary to stress, as von Weizsäcker has done, that the concepts of classical physics play a role in the interpretation of the quantum theory similar to that of the *a priori* forms of perception in the philosophy of Kant. Just as Kant explains the concepts space and time or causality aprioristically, because they already formed the premises of all experiences and could therefore not be considered as the result of experience, so also the concepts of classical physics form an *a priori* basis for experiments in quantum theory, because we can conduct experiments in the atomic field only by using these concepts of classical physics.

It is true that by such a conception we take from Kant's *a priori* a certain pretense of absoluteness which it had in the Kantian philosophy. While Kant could still assume that our *a priori* perceptive forms of space and time must constitute forever an unalterable basis for physics, we now know that this is by no means the case. For example, the complete independence of space and time in nature, which we conceive to be indisputable, does in fact not exist, as shown by very precise observations. Our forms of perception, although *a priori*, do not adapt themselves to the observation of events that occur near the velocity of light, observations that can be made only with very refined technical equipment. Our assertions about space and time must therefore differ, depending on whether we mean our innate *a priori* perceptions or those plans of order existing in nature independent of human observation, in which objective occurrences in the world seem somewhat strained. Similarly, although classical physics is the *a priori* foundation of atomic physics and quantum theory, it is not correct in everything; i.e., there are large areas of phenomena that cannot even be approximately described by the concepts of classical physics.

In these fields of atomic physics much of the earlier intuitive physics is of course lost. Not only the applicability of concepts and laws of that physics, but the whole representation of reality which has been the basis of the exact natural sciences up to the era of today's atomic physics. By the phrase "representation o f reality" we mean here the concept that there are objective phenomena taking place in a definite manner in space and time, whether they are observed or not. In atomic physics the observations can no longer be objectified in such a simple manner; that is, they cannot be referred to something that takes place objectively or in a describable manner in space and time. Here it remains still to be added that the science of nature does not deal with nature itself but in fact with the *science* of nature as man things and describes it.

This does not introduce an element of subjectivity into natural science. We do not by any means pretend that occurrences in the universe depend on our observations, but we point out that natural science stands between nature and man and that we cannot renounce the use of man's intuition or innate conceptions. This character of the quantum theory already makes it difficult to follow wholly the program of materialistic philosophy and to describe the smallest particles of matter, the elementary particles, as the true reality. In the light of the quantum theory these elementary particles are no longer real in the same sense as the objects of daily life, trees or stones, but appear as abstractions derived from the real material of observation in the true sense. But if it becomes impossible to attribute to the elementary particles this existence in the truest sense, it becomes more difficult to consider matter as "the truly real." Because of this, occasional doubts have been voiced in the last few years from the camp of dialectic materialism against today's usual interpretation of the quantum theory.

A fundamentally different interpretation could of course not even have been proposed from that quarter. I should like to mention only one attempt at a new interpretation. It was an attempt to state that the fact that a thing – for example, an electron – belongs to a collectivity, that is, a collection of electrons, is an objective fact that has nothing to do with whether the object has been observed or not and is thus totally independent of the observer. Such a formulation would, however, be justified only if the collectivity really existed. In reality, however, one is deal ing as a rule with only one object, for example, with that one electron, while the whole exists only in our imagination, inasmuch as we consider each experiment repeated as many times as we wish with that one object.

To describe belonging to an only imaginary collective whole as objective fact seems to us hardly possible. We cannot therefore avoid the conclusion that our old representation of reality is no longer applicable to the field of the atom and that we shall find ourselves in very difficult abstractions if we try to describe atoms as that which is truly real. Basically speaking, the same concept of "truly real" has already been discredited by modem physics, and the point of departure of materialistic philosophy must be modified at this point.

III . In the meantime, in the last twenty years the development of atomic physics has led us even further away from the fundamental concepts of materialistic philosophy in the ancient sense. Experiments have shown that the bodies that we must undoubtedly regard as the smallest particles of matter, the so-called elementary particles, are not eternal and unalterable, as was supposed by Democritus, but can be transmuted into one another. Here, of course, we must first state our grounds for describing these elementary particles as the smallest particles of matter. Otherwise it would be possible to believe that these particles are composed of other smaller bodies, which in their turn would be eternal and unalterable. How can the physicist exclude the possibility that the elementary particles themselves are composed of smaller particles, which escaped our observation for one reason or another?

I wish to explain in detail the reply given by modem physics to this question, because it gives prominence to the non-intuitive character of modem atomic physics. To ascertain experimentally if an elementary particle is simple or complex, it is evidently necessary to try to break it up with the strongest means at our disposal. Naturally, there are no knives or tools with which we might attack the elementary particles, the only remaining possibility is to make the particles collide with each other with great energy to see whether they break each other apart.

The large accelerators, which are today in operation in many part of the world or are still under construction, serve this very purpose. One of the largest machines of this kind is, as you now, being constructed by the European organization CERN here in Geneva. With such machines it is possible to accelerate elementary particles to very high velocities (in most cases protons are used) and to make them collide with elementary particles of any other material being used as recipient. The results of such collisions are then studied case by case. Although much experimental material on the results of such collisions must still be collected before we can hope to be completely clear about this branch of physics, it is nevertheless even now possible to say qualitatively what happens in such collision processes.

It has been found that scission can undoubtedly take place. Sometimes a great many particles originate in such a collision, and surprisingly and paradoxically the particles that originate in the collision are no smaller than the elementary particles that were being broken up. They are themselves again elementary particles. This paradox is explained by the fact that, according to the theory of relativity, energy can be converted into mass. The elementary particles to which the accelerators have given a large amount of kinetic energy, with the help of this energy, which can be converted into mass, can generate new elementary particles. Therefore the elementary particles are really the last units of matter, that is, those units into which matter breaks up when maximum forces are used.

We can also express this phenomenon in the following manner: All elementary particles are composed of the same substance, that is, energy. They are the various forms that energy must assume in order to become matter. Here the pair of concepts "content and form," or "sub stance and form," from Aristotle's philosophy reappears. Energy is not only the force that keeps the "all" in continuous motion, it is also – like fire in the philosophy of Heraclitus – the fundamental substance of which the world is made. Matter originates when the substance energy is converted into the form of an elementary particle. According to our knowledge today, there are many such forms. We now know about 25 types of elementary particles, and we have good reason to believe that these forms are all manifestations of certain fundamental structures, that is, consequences of a mathematically expressible fundamental law of which the elementary particles are a solution, in the same manner as the various energy states of the hydrogen atom represent the solution of Schrödinger's differential equation. The elementary particles are therefore the fundamental forms that the substance energy must take in order to become matter, and these basic forms must in some way be determined by a fundamental law expressible in mathematical terms.

This fundamental law sought by present-day physics must satisfy two conditions, both of which follow immediately from experimental knowledge. In the researches on elementary particles, for example, in those performed with large accelerators, so-called rules of selection have been obtained for the transformations that take place following collisions or following radioactive disintegration of particles. These rules, which can be formulated mathematically by means of suitable quantum numbers, are the direct expression of the symmetrical properties inherent in the fundamental equation of matter or its solutions. The fundamental law must therefore contain in some form these observed symmetries or, as we say, represent them mathematically.

Second, if it is conceded that there is such a simple formulation, the fundamental equation of matter must contain, together with the two constants, velocity of light and Planck's action quantum, of which we have already spoken, at least one further similar constant of measure; since the masses of the elementary particles can for purely dimensional reasons follow from the fundamental equation only when, apart from the known constants of measure of which I have already spoken, we introduce at least one more. Observations on atomic nuclei and elementary particles suggest that this third constant of measure should be represented as a universal length, whose order of magnitude should be about 10^{-13} cm.

In the fundamental natural law that determines the form of matter and thus the elementary particles, there must be three fundamental constants. The numerical value of these three constants of measure no longer contains any physical expression. Rather, the numerical value represents only a further expression of the scale by which we wish to measure natural phenomena. The real conceptual core of the fundamental law must, however, be formed by the mathematical properties of symmetry it represents. The most important symmetrical properties of this equation, which is still to be found, are already known on the basis of experience. I should like to list them briefly. First of all the fundamental law must certainly contain the so-called Lorentz group, which can be considered as a representation of the equality of space and time required by the special theory of relativity. Furthermore, the fundamental equation must be at least approximately invariable with respect to a group of transformations that can be described mathematically as the group of unitary transformations of the two complex variables. The physical basis of this property of transformation is a quantum number discovered experimentally more than twenty years ago, which differentiates neutrons from protons and which is now generally known as isospin. In the last few years the research work of Pauli and Gürsey has shown that this quantum number can be represented by the above mathematical transformation. There are furthermore some other group properties, mirror symmetries in space and time, on which, however, we cannot dwell further.

So far, one proposal has been made for the fundamental equation of matter, which satisfies the above-mentioned condition and is moreover very simple. The simplest and most symmetrical nonlinear wave equation for a field operator considered as spinor exactly satisfies the stated conditions. Whether this actually expresses the correct formulation of the natural law can be determined only on the basis of very difficult mathematical analyses in the next few years. I should like to point out here that there are also many physicists who are not so optimistic regarding the simplicity of the mathematical form of the fundamental law. Considering the rather complicated system of observed elementary particles, they suppose rather that there must be a number of different field operators (some say at least four, others at least six) among which there must be a correspondingly complicated system of mathematical relationships. The problem of how this fundamental law can be formulated more or less simply or complexly is not yet decided, and it is to be hoped that the observed data that will be collected in years to come with the aid of powerful accelerators can soon offer a secure basis for the solution of these problems.

Regardless of the ultimate decision, it can even now be said that the final answer will be nearer to the philosophical concepts expressed for example in the *Timaeus* of Plato than to those of the ancient materialists. This fact must not be mistaken for a desire to reject too lightly the ideas of the modem materialism of the nineteenth century, which, since it could work with the whole of the natural science of the seventeenth and eighteenth centuries, comprised much important knowledge that was lacking in ancient natural philosophy. Nevertheless it can not be denied that the elementary particles of present-day physics are related rather more closely to the Platonic bodies than to the atoms of Democritus.

Like the regular elementary bodies of Plato's philosophy, the elementary particles of modem physics are defined by the mathematical conditions of symmetry; they are not eternal and invariable and are therefore hardly what can be called "real" in the true sense of the word. Rather, they are simple representations of those fundamental mathematical structures that are arrived at in the attempts to keep subdividing matter; they represent the content of the fundamental laws of nature. For modem natural science there is no longer in the beginning the material object, but form, mathematical symmetry. And since mathematical structure is in the last analysis an intellectual content, we could say, in the words of Goethe's Faust, "In the beginning was the word"– the logos. To know this logos in all particulars and with complete clarity with respect to the

18

fundamental structure of matter is the task of present-day atomic physics and its unfortunately often complicated apparatus. It seems to me fascinating to think that there is today a struggle in the most diverse countries of the world and with the most powerful means at the disposal of modem technology to solve together problems posed two and a half millennia ago by the Greek philosophers and that we shall perhaps know the answer in a few years or at the latest in a decade or two.

DISCUSSION OF THE LECTURE OF WERNER HEISENBERG

Albert Picot

Werner Heisenberg has expounded to us how nineteenth century science was founded on the ideas of Newton and Descartes. Absolute time, absolute space, absolute causality shut us in and have enclosed the scientists in a relatively narrow area. On the other hand, with the new discoveries of which we spoke last night and above all with the quantum theory and relativity, we have arrived at the concept of which Heisenberg is the chief proponent, the uncertainty principle, a concept that casts doubts on the general theory of causality and on determinism. In a parallel manner in the eighteenth, nineteenth and twentieth centuries, a series of great philosophers affirmed the liberty of man, even independently of science. Three of these appear to us as the proponents of liberty: Kant, Charles Secretan, Karl Jaspers.

And here is my question – a rather critical question, almost indiscreet, since it compels us to ask Professor Heisenberg what his convictions are. Do these philosophers find support in the theory of uncertainty, in the new orientation of science, which recognizes the role of liberty in nature? Is it not strange that a man like Karl Jaspers does not support these discoveries? Is it a new element to prove the liberty of man or is it only a momentary stage in science, which one day will again show causality in quanta? Can we join together, or must we separate, the philosophers who affirm liberty on the basis of philosophy and the scientists who have expounded the uncertainty concept on principles that seem very solid?

Heisenberg

The problem of the relation between uncertainty and liberty has been dealt with too imprecisely and superficially, particularly in the press. It cannot be said that the uncertainty principle opens the door wider to liberty.

We must try to approach the problem of uncertainty and liberty by means of the theory of knowledge, such as Kant also used. The question of what I can do or cannot do is, however, very different from the question what another must do or not do. And there are always many complex questions related to these problems.

Even when we have to do with apparently identical questions, very different replies are obtained; this depends on the way such questions are approached. When we have to do with apparently completely different questions, which often are only different facets of the same problem, very similar answers are sometimes obtained. Summarizing, I do not think that the uncertainty principle has a direct relationship to the concept of liberty. The relation is rather indirect; the introduction of uncertainty into physics has put us on guard against taking too definite a position.

Giacomo Devoto

Before addressing two questions to Professor Heisenberg, I will take the liberty of making a very brief comment on the discussion which has taken place up to now.

1. The relationships to philosophy: It is a very different

thing to say that the progress of physics has repercussions on philosophy and that the progress of physics has given a new aspect to the relationship between science and philosophy. In the first case, we must never forget that philosophy is something that precedes science. For centuries it oscillated between a realistic and idealistic vision of the world. The discoveries of science can influence it in one direction or another, but they are never decisive.

2. I associate myself with Professor Heisenberg's point of view about the concept of liberty. Defining moral liberty by basing it on the uncertainty principle is as absurd as saying: since we cannot put all men into one prison or compel them to live in the same manner from morning till night, we may as well acknowledge their liberty.

Since the uncertainty principle means only one thing – man and science are not in a position to photograph na ture down to its last details – it is ridiculous to try find in it a basis for liberty. A definition of liberty cannot be based on a phenomenon of inability.

I now come to the two questions to Professor Heisenberg that I have already announced.

Since the passage of science from the nineteenth to the twentieth century does not imply a change in philosophical position, is he willing to limit it to a change in definitions?

Up to the nineteenth century, science hoped or pretended to photograph nature. The science of the twentieth century limits itself to describing it. The science of the twentieth century is a language. Being a language, it must arouse the same problems that present themselves in the study of a language. And the fundamental question is the following: In physics there is the relationship between physical facts and mathematical interpretation; in the study of languages, what is there? On the one hand there is historical observation, the history of languages; on the other, the pedagogic application by the grammarian, who tries to establish and describe the conventions that everyone observes in a linguistic context.

Now the second question I ask Professor Heisenberg is the following: Does he accept my suggestion to set up a parallel between physics and the history of languages, of that language which is the new science of which the mathematician is only the grammarian? I know that this definition does not please the mathematician; but nevertheless this is one way of putting the question and above all of ending the discussion between those who believe we can describe facts mathematically and those who do not. Mathematics is a way of describing physical phenomena, just as the rules of grammar are a way of describing a language but are not the language.

Heisenberg

In broad outline I perfectly agree with what you state. It is really possible to say, when recapitulating, that the nineteenth century attempted to photograph nature, while the twentieth century describes nature in a mathematical language. The physicist, however, has realized that, when he believed he was photographing, he was not always doing so.

The physicist in the nineteenth century did not have to discuss philosophy or religion. It was even believed that these doctrines could be kept completely aside and that it would thus be possible to achieve what Professor Devoto calls "a photograph of nature." But it was found that this point of view could not be confirmed experimentally, and very often, when atomic physicists try to photograph nature, they alter its character.

Furthermore, it is observed that the physics of quanta, where uncertainty intervenes, must always be based on deterministic physics. It is hardly possible to do otherwise, and it seems that indeterminateness brings a correction to classical, deterministic physics.

I think it would undoubtedly be useful to study more closely and develop this problem of parallelism between science and language, but I will not do so here. I believe that Professor Devoto is more qualified to attempt it, and perhaps he has already done so. However, we must not forget that the sciences are "between" nature and man.

Ellen Juhnke

I would like to know the reactions of Professor Heisenberg to the idea of Professor Victor von Weizsäcker (ex pressed in his book on creation) that all objective laws of nature already exist, without any contribution by man. The complicated researches by physicists would merely allow them to read as an open book, so to speak, the objective laws governing the organization of nature.

Professor Max Hartmann has said something similar from the standpoint of biology. Starting from the fact that what is called Planck's quantum of action, or the universal constant, is found in mathematical formulae of both the planetary system and the smallest elements of the atoms, he deduces the necessity of a conscious creation by a creator. Among many others, the English philosopher Tomlin presents parallel ideas in his metaphysics, or more precisely, metabiology; and Max Planck has declared, speaking as a philosopher: "Only those who think by halves become atheists, those who go deep with their thoughts and see the marvelous relationships among universal laws recognize a creative power." Or again: "For religion, the idea of God is at the beginning; for science the idea of God is at the end." Is it not an extraordinary thing to see such syntheses being outlined in our atomic and atomized era? My question, therefore, is: What is the opinion of Professor Heisenberg regarding this synthesis between religion and metaphysics and the objectivity of the laws of nature?

Heisenberg

To the first question I reply: The aim of the physicists of the nineteenth and twentieth centuries has certainly remained the same, that is, to find descriptions and objective laws of nature. The difference is that the physicists of the twentieth century have realized that this is not always possible. This difficulty is due to the fact that we are obliged to use a human language for this description. It is evident that, to a certain extent, nature exists independently of man, As Karl von Weizsäcker has said. "Nature existed before man." That is to say, nature certainly existed before man existed, but if nature existed before man, it is not the same as the natural sciences. For example, the concept of "the law of nature" cannot be completely objective, the word "law" being a purely human principle. To reply to the second question, concerning the relationship between religion and science, I would like to cite some ideas of Goethe. In his theory of colors, in particular, Goethe had recognized a certain coherence in the natural orders. He attempted to classify them and placed at the bottom of the scale that which is due purely to chance, then the purely mechanical relationships, then physics, chemistry, biology, psychology and at the top religion, nevertheless realizing that this division was neither exact nor rigorous. The physicists of the twentieth century have become more modest, because they are not certain that it is possible to pass from one field, in which they believe they understand the laws and phenomena, to another which should be adjacent. For example, it is the quantum theory that established the relationship between physics and chemistry, which at one time were completely separate. But when this step was taken, it was realized that it was necessary to change language and to change the orientation of numerous problems. If this passage from physics to chemistry, which are adjacent, is obviously rather difficult, the passage from chemistry to biology will be much more so, and that from biology to psychology will be even more delicate, to say nothing of the passage to religion. The physicists of today realize that the knowledge of the laws in one field will not necessarily permit the passage to another field.

René Schaerer

I shall now ask a first question. A moment ago the name of Descartes was mentioned. I would like to ask Professor Heisenberg why, while passing from today to Aristotle, then to Plato or Democritus, he did not even mention Descartes. Descartes is the thinker who more than any other has made his influence felt today. He dominates philosophy, modem science; he is the first great modem philosopher. The whole history of modem philosophy is generally dated from Descartes; and I note that a thinker as curious about the past as Heisenberg seems to ignore Descartes. Has he any particular reason?

I now pass to the second observation which I feel obliged to make. It seems to me suggestive and curious (I do not believe that it is accidental) that the thinkers to whom Professor Heisenberg refers, first Kant, then Aristotle and Plato, are thinkers who, in their system, assign an important part to finality. Because in the final analysis the elementary geometrical particles of Plato are only the projections of preconceived ideas, known by intuition and all finalized to the good. The "potentia" of Aristotle, which seems to me very different from the uncertainty of Heisenberg (I follow Professor Heisenberg with some difficulty on this point), this potentia is authentically finalized directly. In compensation Professor Heisenberg excludes Democritus, who rejects finality and admits pure mechanism. He does not say anything about Descartes, who is a mechanistic thinker. I therefore permit myself to ask Professor Heisenberg the following question: Can we conclude from

27

this that he tends to admit that there exists a finality, beyond that mechanism known statistically with greater or lesser probability, and that he is therefore in agreement with his great masters, Aristotle and Plato?

Heisenberg

Descartes is certainly at the basis of all present-day philosophy of science, but at the same time he is somewhat at a crossroads. At present it is thought that Descartes was too peremptory in his concepts. It could be said that Descartes' manner of viewing seems comparable to a tennis match, where the ball passes with precision from one court to another, while the manner of thought of a Thomas Aquinas, on the other hand, resembles a game of football, where the whole field is in motion and moves as a whole.

I am furthermore adding a few words on finality. It is evident that, since Newton, causality has served as point of departure. This means that it was thought possible to determine the state of an object or a system, as it will present itself in the future, starting from its previous properties. But, although finality has undergone a slight mitigation in the quantum theory, something of it remains. It follows clearly, in particular from the study of wave mechanics, which explains nearly the whole of chemistry, that a large part of finality survives in an indeterministic concept. If an atom or molecule is disturbed in whatever manner, it will be found after the disturbance that millions of different states are possible, each with a certain probability. But, if I start with a hydrogen atom, I can agitate it in any manner I like and it will always remain hydrogen (as chemical element). This is the characteristic of finality that chemistry brings to physics. The fact that a disturbed hydrogen atom remains hydrogen implies a finality, but a finality of which the causality is not known. In short, this is the fusion of causality and finality that constitutes the basis of modem physics.

René Schaerer

Now I do not understand why, despite his materialism, you are not nearer to Democritus than to Aristotle. I find that you are rather far from Aristotle, as what you have said corresponds rather closely to the idea of Democritus that the atoms undergo or perform a disordered dance in the infinity o f time with infinitely variable velocities. An atom, however, always remains an atom, and here is found the kind of very diminished finality of which you have spoken. On the other hand, I do not see any analogy between the theories developed by you and those of Aristotle. The fact that Democritus is a materialist is not very serious, because this is not the essential, but I find that there are in your system one or two points that seem to come very close to Democritus. Does not the principle of uncertainty perhaps correspond to what Democritus simply called the chance of the game of the atomic dances? And does not what you have just said of the permanence of the hydrogen atom perhaps correspond to what Democritus said about the permanence of each atom, as the atoms cannot be smashed or cut or transformed in any manner? I find you nearer to Democritus and further from Aristotle than you say.

Heisenberg

I resume with some further details of the example of the hydrogen atom. When a simple hydrogen atom is considered and its collision with an electron is studied, a disturbance in the hydrogen atom is observed. The classical physicists believed this collision occurred in a completely analogous manner to that which would have been produced between a planet and comet. In more modem physics the result of this collision is nevertheless not completely pre-dictable, even though it depends on the initial conditions. There exists one probability of finding an electron in the excited hydrogen atom, another of finding the nucleus de-prived of its electron. And these probabilities are fixed and cannot be modified. The hydrogen atom that is found after the collision is, however, no longer exactly what it was before. It is in fact known that, when an interaction contains enough energy, there exists the probability that the hydrogen is not found again, but instead something completely different is found. Several different cases are possible, and these cases are connected among them selves by relations of probability. In fact, what is thus found as a result of an interaction, of any action, is not always objects, but forms – forms of that energy which is the fundamental basic material of modem physics, capable of taking different forms in which we recognize objects.

Umberto Campagnolo

I am wondering whether a physicist can really speak as philosopher and if his considerations will have the precision that a philosopher must observe in his subject.

I believe that there lies the chief danger of the discus-sion of which is to be ascribed, at least from appearances, to the bringing together of scientists and philosophers. The problem a philosopher puts to himself is always of a rad-ically different nature from that which the scientist puts to himself. The scientists assume that there is always the possibility of arriving at quantity and measurement, at calculations and equations. The philosophers, on the con-trary, seek categories and try to link them by means of a process that, if you allow me to use the word, has nothing in common with that of science, that is, by dialectics.

Concluding, I think that we would have much to gain in our discussions by eliminating here the references to philosophy, because philosophical thought is very different from scientific thought. Scientists often tend to imagine philosophy as an extension of science, as a general way of considering their problem from a particular point of view. But I believe that they deceive themselves; it is the philosophers who are and remain responsible for philosophy.

Heisenberg

I completely agree with this way of posing the problem, but I ask Professor Campagnolo if Plato's theory, according to which the ultimate particles of "earth" are cubes and those of "fire" tetrahedrons, is philosophy or science?

Campagnolo

It is possible that Plato used certain ideas of an empirical, poetic character to deify his concepts. At the beginning and end of his life Plato perhaps approached a poetic vision of the world. But in any case poetry is much nearer to science than is philosophy. It is for this reason that I would not object if a scientist considered Lucretius a poet rather than philosopher. If we examine the considerations of Plato on the world, we can be certain that he was far from modem physics and its requirements of equations and numbers. Plato always remains in the field of quality; that of quantity is still foreign to him.

Heisenberg

It is evident that the passage from science to philosophy has given rise to a great many misunderstandings. But I do not believe that it would be useful to try to separate the two fields absolutely and say: Here it is the man of science who is competent, there the philosopher. On the contrary, I believe that it is useful to let the man of science talk philosophy, and the philosopher sometimes science, even at the risk of creating new misunderstandings. The result can be so useful that it is worth running this risk.

Daniel Christoff

The questions that I would like to address to Professor Heisenberg refer to the need of a man of science to adopt philosophical terms which have been defined and discussed at length for a long time.

The first question refers to the "mathematical structure of elementary particles" taken from Plato. I wonder if, taken as a structure of "ideas," it represents an *a priori*? In asking that question I am basing myself on Professor Heisenberg's allusion to the Kantian elements contained in the new theory. What is really *a priori*? Is it the structure of these ideas or the idea of the structure itself?

Heisenberg

These mathematical expressions by means of which we represent particles or phenomena certainly are not *a priori*; but this does not prevent the inclusion of *a priori* concepts in physics. For example, I can imagine a space in which there are no objects, but I cannot prevent myself from thinking that there is a space. It is in this manner that the concept of space becomes *a priori*.

The same can be said of the figures of Plato. Not even here is there an *a priori*, in the sense that Plato would have been able to think that the fundamental elements of "earth" are not cubic but, for example, spherical. Since he could have thought of either a cube or a sphere, there is no absolute *a priori*.

Physicists who deal with the quantum theory are also compelled to use a language taken from ordinary life. We act as if there really were such a thing as an electric current, because, if we forbade all physicists to speak of electric current, they could no longer express their thoughts, they could no longer speak, they would be completely sterile. I consequently believe that it is necessary to take up certain *a priori* forms of classical language, even though their value has perhaps somewhat changed.

Daniel Christoff

May I be allowed to ask another question, which seems to me of very great importance because it is at the center of the questions dealt with by Professor Heisenberg.

He has stated that there exists a relationship between the concept of probability and Aristotle's concept of "potentia." Does this then perhaps mean that everything in the world is virtuality? A virtuality that no doubt fulfills itself constantly but never completely. Because, in correlation with this concept of particles, I am searching for the "act." Is it to be understood that the act is energy? But then, is it not the concrete act that forms each object?

Heisenberg

This is a very difficult problem to deal with. When we consider an electromagnetic wave or a luminous ray that falls on a photographic plate, this luminous wave is the condition that, according to a certain probability, something happens that answers the question: Will a grain of silver form on this plate? The act is the appearance of the grain o f silver; and the luminous wave is the " potentia." "Act" and "potentia" are thus closely connected and, when the incidence of the luminous wave in the act is sought, that is, the grain of silver, this appears as *a priori*. In classical physics, where the phenomena are objective, the traditional language of physics can be used, that is, the language of every day. In modern physics, however, the mathematical structures that are met with indicate the probability of a phenomenon and not the phenomenon itself. And, in this sense, in classical physics, it is the act which is sought in the phenomenon, while "potentia" is to be correlated with the mathematical structures.

Daniel Christoff

Can it be said that this "potentia" has a profound origin?

Heisenberg

To a certain extent, yes.

OUR IMAGE OF MATTER[1]

ERWIN SCHRÖDINGER

1. The crisis – a preview

The title of this lecture was suggested to me (in its French version) by the Committee. I accepted it gladly. However, before I attempt to do justice to it as well as I can, there are two things that I must say in advance. In the first place, the physicist can today no longer make a significant distinction between matter and anything else in his field of research. We no longer consider forces and fields of force as different from matter; we know that these concepts must be merged into one. To be sure, we say that an area of space is free of matter; we call it empty, if there is nothing present except a gravitational field. However, this is not found in reality, because even far out in the universe there is starlight, and that *is* matter. Further more, according to Einstein, gravity and mass[2] are analogous and therefore not separable from each other. Our subject to-

---

[1]Lecture presented September 4, 1952.

[2]EDITOR'S NOTE: This could be misleading. If Schrödinger had written "gravity and matter are analogous," it would be less misleading, especially if one day we discover that matter does not induce the curvature of spacetime, but is itself some state of spacetime. With gravity and mass the situation is different. Gravity is a manifestation of the non-Euclidean geometry (curvature) of spacetime, whereas mass is the measure of *resistance* a particle offers to its (flat- or curved-spacetime) acceleration.

day is therefore the collective image that physics has of spatiotemporal reality.

The second point is: This image of material reality is today more unsettled and uncertain than it has been for a long time. We know a great many interesting details; every week we learn new ones. However, to pick out from the basic concepts those that have been established as fact and to construct from them a clear and easily understood framework of which we could say: this is certainly so, this we all believe today, is impossible. There is a wide spread hypothesis that an objective image of reality in any previously believed interpretation cannot exist. Only the optimists among us (and I consider myself one of them) consider this a philosophical eccentricity, a desperate measure in the face of a great crisis. We hope that the vacillation of concepts and opinions signifies only an in tense process of transformation, which will finally lead to something better than the confused series of formulas that today surround our subject.

It is for me – but also for you, my respected listeners – most annoying that the image of matter that I am expected to build before you does not yet exist, that there are only bare fragments of a more or less partial factual value. A consequence of this is that in this kind of narrative we cannot avoid contradicting at a later point what was said earlier. This is somewhat like Cervantes, who allows Sancho Panza to lose his beloved small donkey on which he is riding, but a few chapters later has forgotten this, so that the good animal is again with us. In order to avoid a similar reproach, I want to draw up a plan of campaign. I shall report later how Max Planck discovered, over fifty years ago, that energy can be transmitted only in indivisible amounts of definite size – the quanta. Since, however, Einstein soon afterwards proved the identity of energy and mass,[3] we are obliged to say to ourselves that

---

[3]EDITOR'S NOTE: Here a double caution is needed. First, al-

the smallest particles of matter, the atoms or corpuscles, which we have known for a very long time and whose existence is demonstrated today in many elegant experiments in a perfectly "palpable" manner, are just quanta of energy and, so to speak, predate Planck's discovery by more than two thousand years. Because of this it seems even more secure. Here a side glance will be offered on the enormous significance of this discreteness or *countability* of all that exists and happens; only thus can Boltzmann's famous statistical theory of the *irreversible* course of nature really be feasible and clearly understandable.

All this is well and good and certainly contains a great deal of truth. But then Sancho Panza's donkey will return – after more than two thousand years. For I must ask you to believe neither that corpuscles are individuals stable in time nor that the transfer of a quantum of energy from one carrier to another takes place in distinct steps. Dis-

---

though experts implicitly assume, when talking about energy, that it is energy of *something* (e.g., of the electromagnetic field), this is not always clear to non-experts who may think (and on many occasions do think) that energy as such exists on its own (as a separate physical entity) without any carrier. Second, one should talk more precisely (for a number of reasons) not about identity, but about *equivalence* of energy and mass. E.g., ordinary particles have *rest* (or proper) mass, whereas photons do not possess such mass (they have only mass *equivalent* to their energy); closely related to this distinction is the *fact* that the mass of ordinary particles increases relativistically when their velocity approaches that of light, whereas photons' velocity is always c and their mass does not increase relativistically. As for over thirty years the concept of relativistic mass has been attacked mostly by over-confident particle physicists, it should be reminded that this concept reflects an *experimental fact* – the *increasing resistance* a particle offers when accelerated to velocities approaching c (*it is this increasing resistance that prevents a particle's velocity from reaching c*) – exactly like ordinary (Newtonian) mass reflects an experimental fact – the *resistance* a particle offers when accelerated. It is ironic that it is particle physicists who turned against relativistic mass given that the increasing resistance of particles (when accelerated to velocities close to c) is most often observed in particle accelerators.

creteness is no doubt involved, but not in the traditional sense of discrete individual particles and certainly not as a discontinuous event. For this would contradict experience from another quarter. The discreteness originates merely as a construct from the laws that govern events. These are still by no means completely understood; but a probably pertinent analogy from the physics of tangible bodies is the manner in which the individual partial tones of a bell result from the finite form of the bell and the laws of elasticity, with which no discontinuity is really connected.

2. Some observations on corpuscles

Let us now begin. The view, already advocated by Leucippus and Democritus in the fifth century B.C., that matter is built up of very small particles, which they called atoms, assumed at the turn of the last century a very definite form as the *corpuscular theory* of matter, entering into interesting details, which became continually clearer and more established during the first decade. Just to outline briefly the fine and fundamentally important individual discoveries made along the way would require two hours of your attention.

The beginning was made by chemistry. Even today there are those who are haunted by the idea that chemistry is the sole and original domain of "atom" and "molecule." From the very hypothetical, somewhat anemic role they played there – the school of Ostwald rejected them flatly – they were for the first time raised to physical reality in the gas theory of Maxwell and Boltzmann. In a gas, these particles are separated by wide spaces, but they are in vigorous motion, they collide again and again, are repelled from each other, and so forth. An accurate pursuit of these processes in thought led first to a full understanding of *all* properties of the gases, elastic and thermal properties, internal friction, thermal conductivity and diffusion, but at the same time to a firm basis for the mechanical theory of heat as a movement of these very small particles, which continually becomes more vigorous with increasing temperature. If this is true, then small bodies that are just visible under the microscope must also be kept in continuous motion by the impact of surrounding molecules, and this motion must increase with rising temperature. This motion of small suspended particles was discovered by Robert Brown (a London physician) as early as 1827, but only in 1905 did Einstein and Smoluchowsky show that it conforms quantitatively to expectations.

In this fruitful period, approximately ten years before and after the turn of the century, so much that was closely related to our subject occurred that it will be difficult to keep it simultaneously in view. Roentgen rays – "light" of very short wave length – and cathode rays – streams of negatively charged particles, the electrons – were discovered. There was the radioactive disintegration of atoms and the radiations emitted during this process – partly streams of particles, precisely those during whose spontaneous emission from the bonding of the atomic nucleus the transformation of one atom into another takes place, and partly "light" of even shorter wave length, which originates during this process. All particles carry an electric charge; the charge is always the very small electric unit charge, measured directly by Millikan, or nearly exactly double or three times it. The mass of these particles could also be measured very accurately, as could that of the atoms themselves.

The determination of the mass of the atoms, so-called mass spectrography, was carried to such fantastic precision by Aston at Cambridge that he was able to answer in the negative a very old question with absolute certainty: they are *not* whole-number multiples of a very small unit. In spite of this, we must imagine these, or more precisely, the heavy but very small positively charged atomic nuclei –

the surrounding negative electrons weigh almost nothing – as built up of a number of hydrogen nuclei (protons), of which of course about half have lost their positive unit charge (neutrons). In a normal carbon atom there are, for example, 6 protons and 6 neutrons. It weighs, in a unit convenient for the comparison

$$\text{Carbon atom} \qquad 12.00053 \pm \ldots$$

$$\text{compared with} \begin{cases} \text{Proton} & 1.00758 \pm \ldots \\ \text{Neutron} & 1.00898 \pm \ldots \end{cases}$$

The unit is $(1.6603 \pm \ldots)\ 10^{-24}$ gram; this, however, does not interest us for the moment. How do we explain the "mass defect," which in our example amounts to almost a tenth of a unit? By the *heat of binding*, which is liberated during the combination of these twelve particles and which is enormously greater in such "nuclear reactions" than in the well-known chemical reactions. In other words, the system loses potential energy, as the twelve particles yield to the energy of attraction by which they are held closely together afterwards. According to Einstein, as already mentioned above, this loss of energy signifies a loss of mass. This is called the packing effect. Moreover, the forces are of course not electric ones – these are indeed repellent – but so-called nuclear forces, which are much stronger but act only at very small distances (about 10^{-13} cm).

3. Wave field and particle: their experimental demonstration

Here you catch me in a contradiction, because I said at the very beginning that today we no longer assume that forces and fields of force are something different from matter. I could easily excuse myself by saying: the field of force of a particle is calculated with the particle. But it is not so. The proved opinion today is rather that every thing – *really everything* – is simultaneously particle and field. Everything has the continuous structure that is familiar to us from the field, as well as the discrete structure familiar to us from the particle. Expressed so generally, this recognition contains certainly a great deal of truth. For it is based on innumerable experimental data. Opinions vary as to details, and of this we shall speak later.

Moreover in the particular case of the field of nuclear force, the particle structure is already fairly well known. The so-called π-mesons, which appear among others during the destruction of an atomic nucleus and which clearly leave behind separate tracks in a photographic emulsion, very probably correspond to it. The nuclear particles themselves, the nucleons – the name under which protons and neutrons are lumped together – which we have been brought up to consider as discrete particles, for their part produce interference patterns in other experiments, when a great many of them are driven against a crystal surface. These patterns leave no doubt that the nucleons also have a continuous wave structure. The difficulty, which is equal in all cases, of combining these two widely different characteristics in one mental picture is still today the major obstacle that makes our image of matter so variable and uncertain.

However, neither the concept of particles nor that of waves is hypothetical. I mentioned in passing the tracks in the photographic emulsion, each indicating the path of a single particle. Known even longer are the tracks in the so-called cloud chamber of C.T.R. Wilson. From these tracks we can observe exceptionally diverse and interesting details regarding the behavior of the single particles, and we can measure them. The bending of their paths in a magnetic field (because they are electrically charged); the mechanical laws in a collision, which takes place approximately as in the case of ideal billiard balls; the destruction of a larger atomic nucleus by means of a direct hit by one of those "cosmic" particles that come from the universe, to be sure in small numbers, but with a tremendous impetus of the single particle, often several million times greater than otherwise observed or artificially produced. There have been efforts in recent times to achieve this at the cost of enormous expenditure, chiefly financed by defense ministries. To be sure it is not possible to shoot anyone with such glancing particles, otherwise we would all now be dead. This study, however, indirectly promises the stepped-up realization of the plan to exterminate mankind, which is close to all our hearts.

It is perhaps as well to say that these interesting observations on single articles, which I cannot wholly describe in my short résumé, can be made only on very rapidly moving particles. The method of the tracks is moreover not the only one. You can easily try out the oldest method yourself, if, one evening in the dark after having got used to the darkness, you examine a luminous number of your wrist watch with a magnifying glass. You will find that it is not uniformly bright, but that it surges and undulates just as the sea often sparkles in the sun. Each scintillating flash is produced by a so-called alpha particle (helium nucleus) emitted by a radioactive atom, which at the same time is transmuted into another one. And this continues in the same manner for a great many years – in a good Swiss watch. Another much used apparatus for the study of cosmic rays is the Geiger-Müller counter which "answers" when it is struck by a single active particle. This is very valuable. Today it is possible with quite familiar methods to amplify this "answer" in such a way that it can release the mechanism of a cloud chamber and the shutter of a camera trained on it at precisely the moment when there is something interesting to photograph in the chamber. This is an important use for these chambers but not the only one. Fifty or more of them are often built in a single apparatus in a complex circuit.

So much for the observation of single particles. Now for the continuous field or wave characteristic. The wave structures of visible light is rather crude (wave length about two-thousandths of a millimeter). It has been very thoroughly investigated for more than a century by means of the effects that occur when two or more or a great many waves cross: diffraction and interference phenomena. The most elegant means for the analysis and measurement of light waves is the diffraction grating, an immense number of fine parallel lines, cut at small equal intervals on a metal mirror, on which light impinging from *one* direction is scattered and recombined in various directions, depending on its wave length.

For the much, much shorter waves of the roentgen spectrum, as well as for "waves of matter" – high velocity streams of particles – the finest diffraction gratings we can cut are somewhat coarse. In the year 1912 Max von Laue discovered the instrument that has since made possible the exact analysis of all these waves, discovered it in naturally grown crystals. This discovery was invaluable, unique of its kind. Not only does it reveal the structure of the crystal – a highly regular arrangement of atoms with the same groups repeated a countless number of times, each at equal intervals in three directions, "length," "width" and "height" – but this discovery was *one* with the use of the periodic fine structure of the crystal for the analysis of waves in place of the diffraction grating. And of course remember this: the natural structure of the crystal comes to our aid exactly at the point where it, that is, the crystalline structure of matter, makes all precision techniques impossible. It would be impossible to cut gratings of such fineness, because the "material" is too coarse. With these crystal gratings the wave nature of roentgen rays was first established and their wave length measured, and later that of the waves of matter, especially electron streams, but also streams of other particles such as neutrons and protons.

4. The quantum theory: Planck, Bohr, de Broglie

I have now told you many things about the structure of matter, but we have not yet spoken of Max Planck and his quantum theory. Everything I have reported so far could very well have taken place even without it. How then did it happen? What is this quantum theory all about? I shall not relate the exact historical events, but instead tell you how the matter seems to us today.

Planck tells us in 1900 – and the essentials are still true today – that he can understand radiation from red-hot iron or a white-hot star, for example the sun, only if this radiation is produced in portions and is transferred from one carrier to another (for example, from atom to atom) in portions. This was astonishing, because this radiation involves energy, which was originally a highly abstract concept, a measure of the reciprocal action or effective action of these very small carriers. The partition into definite portions was highly surprising not only to us, but also to Planck. Five years later Einstein told us that energy possesses mass and that mass is energy, that they are one and the same – and this too has remained true to the present day. Thus our eyes were opened: our dear familiar atoms, corpuscles, particles are Planckian energy quanta. *The carriers of these quanta are themselves quanta.* It makes your head turn. We realize that here is something fundamental that is still not understood. As a matter of fact, our eyes were not opened suddenly. It took twenty or thirty years. And perhaps they are not altogether open even today.

The direct consequence was less far-reaching but still important enough. In 1913 by means of an ingenious and obvious generalization of Planck's statement Niels Bohr taught us to understand the *line spectra* of atoms and molecules and simultaneously the construction of these particles from heavy, positively charged nuclei and light electrons circling around them, each of them carrying a unit negative charge. I must here shirk a detailed explanation of this important transition stage in our knowledge. The fundamental idea is that each of these small systems – atom or molecule – can harbor only definite, *discrete* quantities of energy corresponding to its nature or structure; that during a transition from a higher to a lower "level of energy" it emits the excess as a radiation quantum of quite definite wave length, which is *inversely proportional* to the quantum given up (this was already included in Planck's original hypothesis).

This implies then that a quantum of a given amount manifests itself in a periodic process of quite definite *frequency*, which is *directly proportional* to the quantum (the frequency is equal to the energy quantum, divided by Planck's famous constant h). The really quite obvious deduction, that with a particle mass m, which according to Einstein has an energy mc^2 (where c = the velocity of light), there must be associated a wave process of frequency mc^2/h was first drawn by L. de Broglie in the year 1925, first for the mass m of the electron. Only a few years after this famous doctoral thesis of de Broglie, the theoretical "electron waves" required in his postulate were experimentally demonstrated, in the manner I have already discussed. This was the point of departure for the early recognition, also already mentioned, that everything – *really everything* – is both particle and wave field. Because it is true, isn't it, that, when we hear of a particle of mass m, we will connect with it a wave field of frequency mc^2/h. And when we encounter a wave field of frequency ν (nu), we link with it energy quanta $h\nu$ or, what is the same thing, mass quanta $h\nu/c^2$. Thus de Broglie's dissertation was the point of departure for the complete uncertainty of our conception of matter. In the image of both particles and waves there are elements of truth which we must not abandon. But we do not know how to combine them.

5. Wave field and particle: their theoretical relationship

The *relationship* of the two images in general is known with great clarity and in suprising detail. No one doubts its correctness and general validity. With regard to the combination into a single, concrete, obvious image, opinions are so divided that many consider it entirely impossible. I will now briefly outline the *relationship*. Do not count on forming a uniform, concrete picture; and do not blame my lack of skill in representation nor your own slowness in understanding for your failure – for up to now no one has been successful.

In a wave, two things are distinguishable: *first* the wave surfaces, which form something like a system of onion skins except that they *spread out* in a direction perpendicular to the skin (i.e., to themselves). The analogy in two (instead of three) dimensions is well known to you in the form of the beautiful circles of waves produced by a stone thrown into a pool. The *second*, less evident, are those imaginary lines perpendicular to the wave surfaces in the direction of which the wave progresses at that point, the *wave normals*, which are also called *rays* – to use for all types of waves an expression familiar in the case of light.

Here I hesitate. For what I want and have to say now is certainly important and fundamental, indeed it is even correct, but in a sense that we shall have to restrict so much that it almost contradicts the provisional assertion. This provisional assertion is: *These wave normals or rays correspond to the particle paths.* If you cut a small piece from the wave, about 10 or 20 waves in the line of propagation, and about as big a piece perpendicular to it and destroy ("calm") the rest of the wave, then such a "wave packet" really moves along a ray with exactly the velocity, or at any rate the change of velocity, that is to be expected from a particle of the type in question at the point in question, having regard to any fields of force present.

Even if, in the wave packet or the wave group, we get a sort of clear picture for the particle, which we can work out in many details (e.g., the *impulse* of the particle increases as the wave length decreases; the two are exactly inversely proportional) – nevertheless, for many reasons, we must not take this clear image quite seriously. In the first place it is really somewhat blurred, and the longer the wave length the more blurred it is. Second, there is often not a small packet but an extended wave. Finally, there can be very small "packets" of such a structure that there can be no question of wave surfaces or wave normals – an important case to which I shall return immediately.

The following interpretation seems to me appropriate and representative, because it is extensively supported by experiment: at each point of a regularly progressing wave train there is a *twofold structural relationship of effects*, which it is possible to differentiate as "longitudinal" and "transverse." The transverse structure is that of the wave surfaces and comes to light in experiments on diffraction and interference; the longitudinal structure is that of wave normals and manifests itself in the observation of single particles. Both have been proved convincingly by ingenious experimental designs, each carefully thought out for its special purpose.

All these concepts of longitudinal and perpendicular structure are not exact and absolute because those of the wave surfaces and wave normals are not. They are necessarily lost when the whole wave phenomenon is limited to a small space by the measurement of a single or very few wave lengths. This case is then of very special interest, above all for those waves that, according to de Broglie, constitute the "second nature" of the electron. For them it turns out that this case must occur in the vicinity of a positively charged atomic nucleus so that the wave phenomenon, a kind of stationary vibration, shrinks into a small space for which the real atomic size is found very exactly by calculation.

This size was in fact already fairly well known in an other manner. Stationary water waves of a similar type can be produced in a small washbasin by splashing fairly regularly in the middle with a finger or even by giving the whole basin a slight jolt, so that the water surface rolls to and fro. Here the waves are no longer regularly distributed; but what attracts attention are the *natural frequencies* of these stationary vibrations, which you also observed quite well in the washbasin. For the wave group acting around the atomic nucleus, we can calculate these frequencies, and they are generally found to be exactly equal to the "energy level" of Bohr's theory (which I mentioned briefly earlier) divided by Planck's constant h. The ingenious but nevertheless somewhat artificial assumptions of that theory, as well as those of the older quantum theory, are replaced by a much more natural assumption in de Broglie's wave phenomena. The wave phenomenon forms the real "body" of the atom. It replaces the individual punctiform electrons, which in Bohr's model swarm around the nucleus. In no case can we admit the existence of such punctiform particles within the atom, and if we still think of the nucleus as such, this is an entirely conscious expedient.

Regarding the discovery that the "energy levels" are really only the frequencies of fundamental vibrations, the point that seems to me particularly important is that we can renounce the postulation of a stepwise transition, because two or more natural vibrations can quite readily be excited simultaneously. The discreteness of the *nat-*

ural frequencies suffices completely, as I at least believe, to support the considerations from which Planck started and many similar just as important ones – I mean, in a word, to support the whole of quantum thermodynamics.

6. Quantum steps and identity of particles

The abandonment of the *theory of quantum steps*, which to me personally seems more inadmissible from year to year, has, I admit, important consequences. It really means that the exchange of energy in definitely limited packets is not taken seriously, is not really believed, and is instead replaced by the resonance between vibration frequencies. We have seen, however, that, because of the identity of mass and energy, we must consider the particles themselves as Planckian energy quanta. This is frightening at first. For, with this disbelief, we must also not consider the single particles as a well-defined permanent reality.

There are many other reasons to support the fact that it is not so in reality. In the first place properties have long been attributed to such a particle which contradicts it. From the above fleeting picture of the "wave packet," it is easily possible to infer the famous Heisenberg uncertainty principle, according to which a particle cannot simultaneously be at a specific place and have a clearly determined velocity. Even if this uncertainty were small – and it certainly is not – it follows from it that it is never possible to observe the same particle twice with apodictic certainty.

Another very sound reason to deny to the single particles an identifiable individuality is the following: When we are dealing theoretically with two or more particles of the same type, for example, with the two electrons of a helium atom, we must *efface their identity*, otherwise the results will simply be untrue and not agree with experience. We must count two situations which differ only by an exchange of roles by the electrons not only as equal – that would be obvious – but we must count them as one and the same. If we count them as two equals, it becomes nonsensical. This situation weighs heavily, because it is valid for every kind of particle in whatever number with out exception and because it is directly contrary to every thing that was believed about them in classic atomic theory.

The fact that the individual particle is not a well-defined permanent object of determinable identity or individuality is probably admitted by most theoreticians just as they admit the reason described here for the complete inadequacy of this representation. Despite this, the single particle still plays a part in their representation, deliberations, discussions and writings, and with this I cannot agree. Even more deeply rooted is the idea of stepwise transition, of "quantum steps," at least according to the words and modes of expression, which have become permanently naturalized – of course in very careful technical language, the ordinary meaning of which is very difficult to grasp.

The term "probability of transition" belongs to the permanent vocabulary, for example. But surely we can speak of the probability of an event only if we think that it sometimes actually happens. And in this case, since we refuse to recognize intermediate states, the transition must of course be a sudden one. If it required time, it might well be interrupted in the middle by an unforeseen disturbance. We would then not know where we were. The allegedly exact and fundamental concept would have a gap. In this concept, moreover, probability plays an overriding part. The delicate dilemma, wave versus particle, should resolve itself in such a manner that from the wave field we could calculate merely the probability of encountering a particle of definite properties at a definite position, if we are looking for such a thing.

This interpretation might be quite in keeping with the findings derived from very high-frequency waves ("ultra-rapid particle streams") with special ingeniously thought-out experimental designs. I mean those that I previously cited as observations on single particles. In the *tracks,* which are called particle paths, a *longitudinal* action relationship along the wave normals undoubtedly comes to light. But such a relationship is altogether to be expected in the propagation of a wave front. In any case we have more chance of understanding it from the wave representation than, contrariwise, recognizing the *transverse* action relationship of interference and diffraction from the combined effects of discrete particles, if the reality of waves is denied and we grant them only a kind of *informative* role.

7. Wave identity

Real existence is a term undoubtedly almost hunted down by many philosophical hounds, and its simple, naïve meaning has almost been lost to us. Therefore, I wish to remind you here of something else. We have spoken of the fact that a particle is not individual. The *same* particles are really never observed twice, just as Heraclitus said of the river. We cannot mark an electron – "paint it red" – and not only that, we cannot even think of them as marked, otherwise false results are obtained by incorrect "counting" step by step – for the structure of line spectra, in thermodynamics, and many other cases.

In contrast, it is very easy to imprint an individual structure on a *wave* so that it can be recognized again with complete certainty. Think of light beacons at sea. Following a definite code, each one has a definite prescribed light sequence, for example, three seconds light, five seconds dark, one second light, again five seconds dark and then light again for three seconds, etc. The sailor knows: That is San Sebastian. Similarly for a whistling buoy; only in this case there are sound waves. Or else: you telephone by radiotelephone to a good friend in New York; as soon as he says, "Hullo, hi, this is Edward Meier," you know that his voice has imprinted a structure on the radio wave which has traveled five thousand miles to you and which can be distinguished with certainty from any other. It is not even necessary to go so far. If your wife calls from the garden, "Frank," it is entirely analogous, only in this case they are merely sound waves. The journey is shorter but takes a little longer. Our whole understanding of language is based on imprinted individual wave structure. And what an abundance of details in rapid succession is transmitted to us by the motion picture or the television picture by following the same principle!

Here we are of course dealing with relatively crude wave structures, with which we should perhaps not compare individual particles but rather the tangible bodies surrounding us. And these have nearly all a very pronounced individuality: my old pocket knife, my old felt hat, Zürich's Münster, etc. I have recognized them with certainty a hundred times. But in a remarkable manner, we find the characteristic that, in contrast to the case of particles, we must ascribe individuality to wave phenomena, in the case of elementary waves.

One example must suffice. A limited volume of helium gas, for example, can be considered to consist of many helium atoms or, *instead*, of a superimposition of elementary wave trains of waves of matter. Both points of view lead to the same results for the behavior of the gas on warming, compression, etc. However, we must proceed differently with certain rather intricate "counts," which have to be undertaken in both cases. If we form a mental picture of the particles, that is, the helium atoms, as I have already stated previously, we must not attribute individuality to them. This originally seemed very surprising and has led to lengthy controversies, which have, however, long since been settled. On the other hand, in the second manner of consideration, which, instead of particles, imagines wave trains of matter, each is accorded a specifiable structure, different from that of any other. To be sure, there exist many pairs, which are so similar to each other that they can exchange their parts, without our noticing it from outside the body of gas. If we tried to *count* the *very many* similar conditions that arise in this way as only one, we would obtain something quite incorrect.

8. Conclusion

You may be surprised that despite everything I finally have put forward, which in reality no one denies, the closely related concepts of the *quantum steps* and the individual particle have not disappeared either from the vocabulary or from the imagination of the physicist. You will find the explanation, if you consider that the interpretation, which we ultimately reached and toward which we steered in about the last third of my lecture of today, invalidates or at least casts doubt on the real significance of many details of the structure of matter that I put forward in the first two-thirds. However, I was able to use – without unbearable verbosity I could not have helped using – a language that I really do not consider appropriate. How can we state the weight of a carbon nucleus and a hydrogen nucleus to an accuracy of several decimal places and establish that the former is slightly lighter than the twelve hydrogen nuclei combined in it, if we do not provisionally accept the point of view that these particles are something quite concretely real? This is so much more convenient and evident that we cannot abandon it, just as the chemist cannot renounce his valence bonds, although he knows quite well that they are a drastic simplification of a quite involved wave-mechanical situation.

If you ask me: Now, really, what *are* these particles, these atoms and molecules? I should have to admit that I know as little about it as where Sancho Panza's second donkey came from. However, to say something, even if not something momentous: They can at the most perhaps be thought of as more or less temporary creations within the wave field, whose structure and structural variety, in the widest sense of the word, are so clearly and sharply determined by means of wave laws as they recur always in the same manner, that much takes place *as if* they were a permanent material reality. The very exactly specifiable mass and charge of the particles we must thus consider as Gestalt elements determined by the wave laws. The conservation of charges and mass on a large scale must be considered as a statistical effect, based on the "law of the large numbers."

Reflections of a European Man of Science[1]

Max Born

1. Introduction

When I thought what I could say on physics in Europe or on European physics to interest an audience of "non physicists," I found myself confronted with an extremely difficult task and I almost gave up the attempt. For the natural sciences, and above all physics, are by their very essence international. They cannot be bound by the limits of particular countries or continents. There was only one solution: to speak to you not of physics or its history – the greater part of which has taken place in Europe – but of the history of the world as seen by a physicist and of the part played by Europe.

I shall use a method which we physicists have been reproached for employing but which has given proof of remarkable achievements in the realm of science. It is the method of simplifying thought by stressing one aspect only of the facts. I wish to see the varicolored image of history through colored glasses which let through only one tone, but a fundamental one; in this way one gains in clarity what one loses in richness.

---

[1]Lecture delivered September 5, 1957.

2. Energy, a historical factor

Let us examine Europe from the point of view of its technological evolution. I hold that it is essential to consider the energy at the disposal of man as one of the preponderant factors in the history of humanity. Thus one can divide history into two great periods: the first extending from Adam to the present, the second beginning in our time and leading towards the future. The crucial moment is the transition from the use of solar energy to the exploitation of sources of energy of purely terrestrial origin. I consider that the change which is taking place before our eyes is a phenomenon of immeasurable significance which is in no way comparable to anything that has happened up to to day. It is fitting that this evolution should be studied in the setting of our European discussions, for it has been principally a European phenomenon. I am going to demonstrate this fact by first establishing the physicotechnical bases which will provide a foundation for our under standing of the course o f history.

On earth all energy is ultimately based on the processes which take place in the nucleus of the atom.

Life is sustained on earth by the sun's radiation, and this radiation is the expression of the energy produced by the nuclear processes which are located in the sun.

Fifteen years ago or so, man had no energy at his disposal except that of the sun, stored by the atmosphere and the plants. From the point of view of energy, man was still in the first period of history. This period is divided into three energy chapters which are clearly defined: the first, from the most distant ages up to the fire arm; the second, from then up to the steam engine; and the third up to the first atomic reactor in 1942, a critical year which marks the opening of a new era.

3. Structure of the atom and atomic energy

First of all I would like to examine the question briefly from the point of view of physics. Today the general ou lines of atomic theory are common knowledge. Everyone knows that matter is composed of atoms with a diameter measuring approximately one ten millionth of a millimeter. But it is a mistake that the atom should bear this name, of Greek origin, because it is not indivisible. It is made up of a very small nucleus charged with positive electricity, and it is surrounded by a cloud of negative electrons, to such a number that the whole is electrically neutral. The mass of the electron is about 1,800 times smaller than that of the lightest nucleus, namely, the hydrogen nucleus. The latter is called the proton, and its charge is the same, apart from its sign, as that of the electron. The nuclei of other atoms are compact agglomerations of protons and neutrons; neutrons are uncharged particles with a mass almost equal to that of the protons. These two types of particles are termed nucleons. Atoms whose nuclei have the same number of protons possess an identical cloud of electrons, and that is why they have a similar external action, even if the number of neutrons contained in their respective nuclei differs and their masses are consequently different. These chemically identical atoms are called isotopes; the chemical elements are mixtures of isotopes.

All the physical and chemical properties of matter are conditioned by phenomena which are localized in the clouds of electrons; all the radioactive processes, both natural and artificial, are phenomena which are located in the nuclei of atoms.

The nuclei are protected by their clouds of electrons. That is why it was not until a rather late period, only fifty years ago, that the physicists were able to get at them. The diameter of the clouds of electrons, in the order of size, is about 10,000 times greater that that of the nuclei. In contrast, the energy which binds an electron to its cloud is much less (from 100,000 to a million times smaller) than the binding energy that retains a nucleon in the nucleus.

I have often been asked why it is precisely the smallest particles which carry the greatest energy. A detailed analysis of this phenomenon would take us too far. Perhaps it will be sufficient to refer to Newton's law of attraction, which is well known to everyone, according to which two masses, as for example the sun and a planet, attract each other with a force in inverse proportion to the square of the distance between them. The work which is necessary to move the two bodies from their initial position and to take them to a distance so that the force becomes negligible is called the binding energy in the initial position. In the case of Newton's law, it is in inverse proportion to the square of the distance in a given position. If the earth were placed in an orbit half as near again to the sun as it is now, it would be linked to the sun by an energy four times as strong.

According to Coulomb, these same laws govern the forces of attraction and repulsion between electrically charged particles. As the protons and electrons are charged, it can immediately be seen that the contribution of electric forces to the binding energy must be infinitely greater for the protons, which are closely concentrated in the nucleus, than for the cloud of electrons, which is at a distance from the nucleus.

But that is not all. The protons are all charged in the same way (positively), and therefore they repel each other. For the formation of the nucleus to be possible, there must exist forces of another nature and with little power which cause an attraction between the nucleons.

Now here is the reason why I have spoken of these forces in connection with the theme "Europe." The experiments which have made it possible for us to explain the structure of the atom have been carried out by Europeans and Americans. The theoretical interpretation, that is to say the feat of reducing these observations to simple fundamental laws, has been almost entirely the work of Europeans. It is almost impossible to quote names without giving the history of modem physics. I will name two workers only: Rutherford, who did experimental research on the structure of the atom – nucleus and cloud – and Niels Bohr, who established the theory of the cloud of electrons and deduced from known natural constants the factor, mentioned above, of a size of the order of 10,000. When Bohr was seeking to establish his quantitative theory of the structure of the atom, he turned to the two great guiding ideas of modem physics: Einstein's theory of relativity and Planck's quantum theory. Both are typically representative of European thought, and their significance reaches far beyond the natural sciences, entering as they do the field of philosophy.

Is it not characteristic of our era that the interpretation of nuclear forces with slight power, which I spoke of a moment ago, is due to a non-European personality, the Japanese Yukawa (1935), whose work is based on the two great theories quoted above? Yukawa opened broad and completely new perspectives to physics by revealing the existence of short-lived particles, called mesons, with a mass between that of the electron and proton. Since that time several kinds of particles of that type have in fact been discovered. The study of them – which will probably solve the mystery of the origins of matter – will probably be the most important task that the physics of the future will have to confront.

It is no exaggeration to say that the "most European" of all creations of the human mind, apart from polyphonic music, is theoretical physics, which has no equivalent in other civilizations. Yukawa abolished this monopoly.

4. Nuclear transformation and solar radiation

After these incursions into the realm of physics, let us return to our historical considerations on energy.

When I studied physics and astronomy fifty-five years ago, the source of the energy which is radiated constantly from the stars was considered inexplicable. All the known processes, for example the transformation of the energy of gravitation into heat by contraction (as put forward by Helmholtz), could not explain everything. Radioactivity had just been discovered. It was very soon supposed that the radioactive processes, that is to say the nuclear transformations inside the stars, could produce the necessary energy. However, it was only in 1938 that Bethe and von Weizsäcker arrived independently at the correct solution.

The small nuclei are unstable in the sense that they have the tendency to fuse into larger ones, liberating energy. For example, the helium nucleus, the second in the order of atomic weights, is composed of two protons and two neutrons. But it is so improbable that these four infinitesimal particles should be capable of uniting at a given moment that this does not happen directly even in the most compressed matter at the center of a star. It is possible only by a complicated process, by way of other particles which act as chemical catalysts. Here are the conclusions reached by the research workers cited above.

The sun, like all the stars, shines because of this process of fusion. A small part of the sun's radiation reaches the earth and supplies the energy to which we owe our meteorological conditions and the possibility of life on our planet. The heat produced by the radiation keeps the water of the oceans liquid, except in the polar regions, and sets in motion the meteorological cycle: sea – cloud – rain – river – sea, etc.

The plants absorb and assimilate certain short waves

of radiation. What takes place then is a complicated photo chemical process, that is to say a regrouping of the electrons in the clouds enveloping groups of atoms. The energy thus transformed is, per atom, infinitesimal in comparison with that produced by each fusion in the interior of the sun; the energy is reduced to almost nothing by its passage through the sun and its propagation in space. However, it is this chemico-vegetal energy which maintains all life on earth and with which man has been contented up to our day.

5. First chapter of the chemical era: the natural age

The sources of energy man had at his disposal in the first chapter of his history, which one could entitle the "natural state," were his own muscle power and that of his domestic animals. He added to these by a modest contribution from the meteorological cycle: water mills and windmills for work and sailing ships for transport. From the point of view of the natural sciences, it is just this source of natural muscular energy which has been understood least. It consists of a transformation of chemical energy (that is, a regrouping of the electron layers of groups of atoms) in a rough mechanical movement, without a noticeable rise in temperature. Our laboratories are acquainted with processes of this kind only in primitive forms of apparatus, such as electric batteries. What takes place in the organism is extraordinarily complex and subtle. An eminent biologist told me recently that in his opinion a technical imitation of this process would be the equivalent of the synthetic production of living substance.

In the natural state man lives, from the point of view of energy, not on his capital but on his revenue, and this revenue – solar radiation – is distributed everywhere, although irregularly, according to zones.

Given the universality of natural conditions of life on the earth, these conditions are of secondary importance in the evolution of history, while other factors are decisive: geography, national character and personality. Thus historians generally consider the problem of energy as a known factor and turn their attention to other things.

This attitude can be justified in the setting of the natural period, but it becomes erroneous, and even dangerous, in our own age. Great upheavals have taken place, and if we consider them as an appendix to the chapter on economic conditions or cultural questions, we shall not evaluate them adequately.

During this period Europe did not play a role which distinguished her particularly from the other continents: she, too, had her wars and peace treaties, her princes and heroes, her constitutions and revolutions, her philosophies, her religions, her arts and sciences, and all that these entail. But even in this epoch one phenomenon set Europe apart: the appearance of the Greeks, who conceived of free and independent thought. The Greeks, without any intention of making an immediate and practical use of it, sought to discover the nature of the world and were the first to acquire profound knowledge in mathematics and the natural sciences. Certainly this knowledge was subsequently forgotten, but it was rediscovered when the true flowering of Europe began a thousand years later.

6. Second chapter of the chemical era: the age of transition

Gunpowder is said to have been invented in China, where, it seems, it was mainly used for the pleasurable spectacle of fireworks.

When it appeared in Europe in the twelfth or thirteenth century, it was immediately used for warlike ends.

I place it at the head of the second chapter of the chemical era because it represents the first use of chemical energy not stored in living muscles. I consider it as a symbol of the European spirit as it has manifested itself since then, characterized by perspicacity in the spirit of invention, by the need for expansion which, in spite of the teachings of Christianity – sometimes even in its name – has not hesitated before any violence.

This is a period of transition and tumultuous development. It is difficult to distinguish the spiritual from the material elements, for if the religious and philosophical traditions had not been surmounted, the prodigious progress in scientific research which came about during the course of these centuries would have been inconceivable. On the other hand, the success of scientific research contributed to the break with outdated dogmas.

With the great voyages of discovery the roundness of the earth became a reality, and the European, with his cannons, became the master of vast regions of the globe. He imagined himself the master of the world, for he believed that the earth was the center of the cosmos. But Copernicus dethroned the earth and made it merely one planet among many others. This hardly disturbed the self-sufficiency of the European; in science he found a compensation for the loss of his imaginary superiority. Did science not offer him the solution to the riddle of the heavens and, before long, to that of terrestrial nature? Mechanics was born from the study of planetary movement, which in turn gave a powerful impulse to physics. From the medieval mysticism of alchemy arose the exact science of chemistry. At the end of the eighteenth century, after the period of preparation, the steam engine was invented.

During all these centuries of discoveries and inventions the sources of energy had remained the same as at the outset of history. All work continued to be done by the muscles of men and domestic animals aided by water mills and windmills.

7. Third chapter of the chemical eras: the age of fossil fuels

Then a radical change came about. The steam engine depended on coal, which was used as a fuel in England when the ancient forests had been decimated, especially to satisfy the insatiable demand for wood for naval construction. The first steam engines were used to pump water in the coal mines. They themselves consumed coal in ever increasing quantities and in this way lived on the capital of energy deposited in the earth by the sun during the course of hundreds of millions of years, in the form of successive generations of forests which had become decomposed, buried and transformed into coal. The annual production of mechanical energy then rose rapidly and transformed human conditions in Western Europe. The sociologists speak of the industrial revolution, which is an inexact term for what was in reality a revolution in the exploitation of energy. All that followed was only a result of this transformation. Until then man in his work had only one "man power" at his disposal. From then on he was endowed with a larger and larger number of "horse power." This number increases from year to year and differs from one country to another. It is greatest by far in the United States of America: at the present a workman there has an average of about 40 horse-power at his command. This is accompanied by an increase in the production of goods and a rise in the standard of living.

At first, however, this new wealth went into the pockets of employers, while the situation of the masses deteriorated and long years and political revolutions were necessary before general well-being improved. But it is not for me to deal with these modifications in the social structure. I would just like to point out some characteristics of this period.

The first relates to the reciprocal influence between technological advance and science. The invention of the steam engine took place before the theory explaining the principle on which it functions was developed. Even the notion of energy which now provides the explanation of this machine and which I take as a basis with the liberty and familiarity proper to the physicist in order to give an over all picture of the history of humanity, this notion was not developed systematically under the heading of the principle of equivalence between heat and mechanical energy until fifty years after that invention. Later this theory, completed by a second fundamental concept, that of entropy, contributed greatly to the perfection of the steam engine. These reciprocal exchanges have continued with technological and scientific progress in all fields of research and industry,where they have proved their worth.

As my second point, I should like to mention a few of the principal examples of this reciprocal action in the realm of electricity and chemistry. Thanks to electrical technology, energy has become transportable and has been made into a commodity; chemistry, on the other hand, has freed man from his dependence on natural materials. It is no longer possible to enumerate the innovations that the world has seen since then. Seventy years ago, when I was a child, bicycles were not yet in common use. Today we have supersonic airplanes. I am still filled with wonder when I realize that this technical era is still only twice my own age and that the most amazing achievements appear in its second half, that which I have lived through. The most surprising of all is perhaps the triumph of medicine, illustrated by the fact, among others, that it has doubled the average length of human life. In 1900, when my father died at the age of fifty, he was considered quite old; I am now in my seventy-fifth year and, as you can see, still flourishing! All the same, I cannot say that I feel at home in the present-day world.

The third point that I want to consider regards liquid fuel: petroleum. It is extracted from the earth in great quantities and has therefore become an important factor in the economic and political struggle for world supremacy. But if our oil deposits had not existed, there would probably not have been fewer buses and planes, because the previous generation had already learned how to extract liquid fuel from coal.

What a really fantastic age these last hundred and fifty years have been, during which the oil and coal deposits have been exploited! Considered from a detached point of view, it has been grandiose both for its conquests and for the evident absurdity of its undertakings. It is obvious that a given reserve of good things must eventually come to an end if one draws on it constantly, and this end will arrive all the more quickly the more one draws on it. Europeans, including the Russians and Americans of European origin, have lived from day to day without an eye to the future. They extended and consolidated their domination over other peoples, which they had been able to establish with the help of their cannons. After the Napoleonic wars they were so busy with their colonial enterprises that peace reigned among them. The mid nineteenth century was one of the rare periods of prolonged peace in Europe. But the European nations soon began to quarrel again, in Europe itself, over the question of booty from their colonial possessions and over old problems of hegemony and frontiers. Armies gradually became mechanized and industrialized like all other sec tors of life.

The horrors of war consequently increased, and we have seen the result: Europe ravaged by two world wars and deprived of her political dominance in the world – although certain nations refuse to believe it yet. The two great powers of today, the United States of America and the Soviet Union, continue the old, dangerous game of power politics, now intensified by their ideological opposition – liberal capitalism versus totalitarian communism – an opposition which resembles the religious struggles of past centuries for the fanaticism of the two adversaries, both convinced that they have right entirely on their side.

To me the most striking characteristic of the age whose end we are witnessing is the irresponsible way in which humanity is exploiting the fossil fuels – coal and petroleum – to the bitter end, for these are the very sources of its power and greatness. The mighty development brought about by the exploitation of these sources of energy gave rise to an optimism, a faith which was unwilling to recognize any limit to progress. In Europe this faith has been profoundly shaken by the two world wars, but not in America or Russia. And yet fifteen years ago this belief already had no foundation. Coal and petroleum would become exhausted, all the more quickly as the population of the globe increased. In Europe, America and Australia progress in medicine and hygiene was responsible for demographic growth. Peoples from other regions of the earth in ever growing number, above all the immense populations of China and India, aspired to a higher standard of living and began to become industrialized. More and more coal deposits and oil deposits were discovered, and it became unnecessary to worry over the decades – perhaps even centuries – to come.

But the naturalist regards present-day civilization as a brief interlude at the end of a long period – half a million years – in the history of man, the latter period being in its turn only a minute interlude in the endless millions of years of the evolution of life on earth. The naturalist must, in fact, have the right to apply another measure and observe that humanity's faith in its long domination over the world rested, until not long ago, on a very fragile basis.

67

8. First chapter of the atomic era: nuclear fission

To explain this title it is sufficient to remember that the great physicist Lord Rutherford, the father of contemporary nuclear research who discovered the nucleus of the atom, was still convinced, up to his death, in 1937, that it would never be possible to utilize the immense reserves of energy accumulated in the nucleus. He was wrong. Two years later, in 1939, one of his pupils, who like Rutherford dedicated himself to disinterested research as understood by the Greeks, the German Otto Hahn, with his collaborator Strassman, made the decisive experiment, without being aware of its full significance. It is probable that it would have been a number of years before this discovery was made use of for technical ends if the Second World War had not broken out and speeded up research, in the manner of a chemical catalyst. These events are so well known today that it is unnecessary to repeat them. I should just like to make two comments on the subject.

The first regards the physical nature of the new source of energy. It does not consist of a process of fusion caused by solar energy, but one of division or fission of heavy nuclei. The principle is easy to understand. I have explained previously that the cohesion of nucleons in the nucleus cannot be explained by electric forces because, first, all the protons are charged (positively) and therefore repel each other, and because, second, the neutrons in the nucleon are not affected by electric forces. I have related how Yukawa deduced from fundamental principles of modem physics the existence of new kinds of forces with a short range and connected their presence with a new kind of particle, mesons. These forces revealed by Yukawa affect only the immediate surroundings, while the force of electric repulsion can reach greater distances and act also on nucleons further removed. Thus it can be understood that for the big nuclei the repelling electric force, despite its relative weakness, takes the upper hand and prevents nuclei above a certain size from being stable. This is what happens with the uranium element containing 92 protons. It has been established that one of its isotopes – not the most common, which has 146 neutrons – but the one that is considerably rarer and possesses 143 neutrons, becomes unstable when it absorbs an outside neutron and then divides into two approximately equal parts, at the same time liberating powerful energy. Through this, several isolated neutrons are expelled which will provoke in their turn the fission of other uranium atoms. In this way a cascade of fissions is formed, or a chain reaction. This is the phenomenon which is made use of in the uranium reactors to produce energy and which was the basis of the first atomic bomb.

My second observation concerns the role played by Europe on this occasion. The first decisive steps, that is to say the discovery of nuclear fission itself, its theoretical interpretation and the possibility of provoking a chain reaction, have been, without exception, the work of European research workers in Europe or of European origin in the United States; its entire physicotechnical accomplishment has been the fruit of American determination and organization.

The tragic side of this discovery was the first application of this new power to use as a weapon of unimaginable strength. We shall speak of this again later.

Shortly after the war the production of energy began and today several uranium reactors are already function ing in various countries, and numerous others are under construction.

The raw material necessary for the production of this energy, uranium and thorium, like coal and petroleum, exists only in limited quantity, sufficient, however, to delay by several centuries the catastrophe that the exhaustion of the sources of energy would have created.

England, the country where the steam engine was invented, was one of the industrial nations most menaced by the exhaustion of coal reserves. She is now to the fore front in the development of uranium reactors, which she hopes will allow her to maintain her position in the world. Many countries which are lacking in coal and petroleum and, therefore, industry, are today thinking of becoming industrialized with the help of factories using uranium motive power. And already the next stage, when practically unlimited quantities of nuclear raw material will be put at our disposal, is taking shape.

9. Second chapter of the atomic era: nuclear fusion

Man has already succeeded in reproducing on earth the fusion process (by raising the number of nuclei in the helium nucleus to four) which produces the energy of the stars. A uranium bomb was used as a detonator; the prodigious pressures and temperatures which are produced following an explosion by fission are sufficient to set off the fusion reaction. Once more, as in the cases of gunpowder and the uranium bomb, war, or at least the preparations for war, have been the indirect causes of technological progress.

The history of the hydrogen bomb is well known and I do not need to go over it again. The bomb is the decisive factor in the struggle for supremacy between the great world powers, the United States and the Soviet Union. Europe took no part in this struggle until Great Britain began to produce H-bombs. At the outset it appeared to be a purely diabolic invention because no means of slowing down the fusion reaction was known. But very soon methods were discovered which before long will probably allow man to master the so-called thermo-nuclear processes. If he succeeds, humanity will be relieved of all cares regarding its energy reserves for a period of time which will no longer be counted by centuries but by geological periods. In fact, the raw material is a hydrogen isotope which one can extract from sea water, and the oceans will last just as long as the human race on this earth.

In this way man would find himself once again in a healthy position; he would live on reserves of cosmic energy, which are almost as inexhaustible as solar energy. But this new state would differ from the natural state – which prevailed during the first period of the history of energy – in three essential points:

First, it would be an artificial state which could be maintained only with the help of the permanent use of the most intricate technical means and of international collaboration.

Second, judicious employment of these means would assure a state of material wealth; the energy at the disposal of the workman would no longer be the small quantity supplied by his muscles and drawn from his daily food, but an unlimited quantity which he could summon up, as if by magic, thanks to his brain, to science, technology and organization.

Third, it would be an extremely unstable state with inherent dangers of a magnitude quite different from those of the pretechnical period. The catastrophes provoked by the wars and revolutions of the past concerned or wrought havoc upon only limited regions; in the future a political catastrophe would mean the self-destruction of civilization, perhaps of the whole of humanity and perhaps even of life on earth.

10. Perspectives

Let us sum up now. The intellectual and practical achievements of Europe, have rendered man independent of the slender revenue o f solar energy that nature has allotted to him. The European discovered the solar energy accumulated during the course of time – the fossil fuels; and seduced by the lure of wealth he has squandered it without restraint to develop a civilization embracing the whole of humanity. Yet in his pursuit of material profit, he has not completely forgotten the Greek spirit which had given him the initial impulse; and he has continued to pursue disinterested research, thus enabling us to avoid the destitution to which the exhaustion of fossil fuels would have led us: our salvation will come from the utilization of nuclear energy of cosmic origin which is present on the earth itself.

Just as Prometheus had to atone for having stolen fire from the gods and bringing it to man, a curse lies on the achievements of contemporary man for having lit cosmic fire on earth. The atomic era has, in fact, opened with terrible destruction and wholesale massacres, and never will the shadow which the very name of the atom bomb casts over the joy and hope of life be dispelled.

That is the point the scientists have reached. Now it is up to us all, without exception, and no longer only to the politicians, to avoid a cataclysm. We physicists have as our duty to inform and warn statesmen and to do all we can to influence their decisions. That is the aim of this attempt at a scientific explanation of the history of the world and the part that Europe has played in it. The greatest danger for the future comes from those who re fuse to recognize that the nascent era differs entirely from the past. I have enumerated its three essential characteristics: The first, expenditure on technology, is a burden. The second, general material well-being, is an attractive goal – so long as it does not become an end in itself. The third, the atom bomb, is a monstrous danger. The question, then, is to know if we cannot have the well-being without the burden and the danger or – if the burden is inevitable – at least without the danger. What is tragic about our situation is that it seems this would have been possible had the chain of events been different. Thanks to physicotechnical means we could have been content with solar energy without turning to terrestrial nuclear energy. Hydraulic power, which first springs to mind, would, however, have been insufficient; if all the possible hydro electric systems were put to use, only a slender percentage of needs would be covered. Exploitation of the wind is too uncertain. Utilization of the tides is under study and promises appreciable results. On the other hand, the direct conversion of the sun's rays into electric current with the aid of thermoelements is a serious possibility which is being studied particularly by Russian scientists.

I shall quote some figures which I have obtained from a publication by the Russian physicist Joffe: the solar energy sent to the earth in the course of a single day cor responds approximately to the sum of all the quantities of energy which have been accumulated on earth under every form – coal, petroleum, water – since the beginnings of time. This demonstrates that the poverty of the natural era was due not to a limited quantity of radiation but to the absolutely minimum useful effect of meteorological phenomena and vegetable growth. Today the output of thermoelectric installations, even of small steam engines, is from 8 to 10%. However, to meet the whole of the world's needs, a large surface would be necessary, equivalent to a square with sides 30 miles long, in a desert constantly bathed in sunlight.

But even if these projects were realized, it would in no way modify the tragic situation we have arrived at by dropping the Hiroshima and Nagasaki bombs. War and power have presided over the birth of the new era. We have abused a gift of destiny to kill and destroy. A curse will always lie upon us for this act of profanation.

We shall not succeed in escaping from this situation

with out-of-date political concepts. It is often said that, when the crossbow and gunpowder were invented, the end of the world was already being prophesied; we have survived them as we have dynamite, aerial torpedoes and napalm; in the same way we could survive A- and H-bombs – or at least some of us – if we took the necessary protective measures, by hiding in underground cities and taking other similar precautions. Those who speak like this seem to me to be madmen. We are not moles; we take pleasure in the beauties of life, in the sun and the landscapes in bloom which surround us. We cannot escape from the danger which threatens us if we do not radically change our way of thinking. But the difficulties are great, for the world has never been in such upheaval as it is today. The peoples of Asia and Africa wish to rid them selves of the yoke of colonialism and reject the influence of Europe. Nationalism, religious contrasts, racial tensions, ideological conflicts, bringing into opposition, for example, totalitarian communism and liberal capitalism, are more acute than ever. But these differences can never be resolved by the old methods of force. A new world war would mean total annihilation.

Europe provided the original impulse for the achievements of our time with its inventions and discoveries; but these achievements of the spirit have been directed exclusively towards material progress. It seems to me that Europe must take the lead once again in the ethical and political evolution of humanity, and in order to succeed, it must begin by achieving its own unification.

In my capacity as a physicist I am especially interested in European institutions which deal directly with atomic research, such as Euratom and CERN, whose laboratories are located at Geneva. The colossal dimensions of the machines which are installed there show that the smallest products of nature supply the greatest amounts of energy, and consequently their study entails great expense in the experimental field. It is a good sign of the solidarity between the nations of Western Europe that they have associated in order to carry out an enterprise which would have been beyond the possibilities of any one of them.

Physics in itself is not only a factor of material progress but also an element in the spiritual evolution of man. In the final analysis, the opposition between East and West which is preoccupying the world today is based on philosophical opinions and on ways of life which are subject to the influence of the natural sciences. Marxism teaches that the communist economy is a historical necessity and derives its fanaticism from this belief. This idea comes from physical determinism, which itself arises from Newton's celestial mechanics. But, in fact, physics abandoned this theory about thirty years ago. Instead it has worked out a statistical interpretation of natural laws which corresponds better to reality and in the light of which the communist belief that Marxist predictions will necessarily be realized appears grotesque. American thought, for its part, is at the mercy of a superficial pragmatism which confuses truth and utility. I cannot adhere to it. I believe, for example, that the laws of nuclear physics contain a large part of truth but that only the future will be able to tell us if they will ultimately be useful to humanity or if they will bring only death and destruction.

Europe is not bound to one or the other of these extreme and absurd doctrines. We believe that an intermediate and reasonable solution must exist and that it is useless to jeopardize the existence of civilized humanity to secure the triumph of a doctrine or an economic system. For my part, I believe that genocide and war are to be condemned whatever the circumstances are, and I ask that, in the future, politics should no longer have recourse to these means. But as I have set myself the task of considering historical problems as a physicist, it would exceed my competence even more to speak of moral philosophy or even of theology. But I would like to add, in conclusion, that the ethical problems raised by the prodigious increase in the power at man's disposal absorb me just as much as, if not more than, the scientific and political problems.

THE METHODS AND LIMITS OF SCIENTIFIC KNOWLEDGE[1]

PIERRE AUGER

Having been invited to speak to you as a physicist, I believe I owe it to you to assume a scientific attitude from the beginning. I shall therefore summarize the situation I am in as follows: The occasion that brings us all together bears the general title, *International Meetings.* The subject of tonight's talk has to do with scientific knowledge, and moreover the general theme of these Meetings is *Man and Science.* From these premises it is clear that you expect of science that it should, in the person of one of its representatives, come forth and extend its hand to man, that it should in some way become humanized. I shall endeavor to meet this expectation, and at the same time I shall ask this man who faces me – the nonscientist – to be good enough also to take a few steps toward science.

But a word of caution: we must be logical. Science is *of* man – that no one denies. And yet here we find man – or at least a fraction of the human species that considers itself sufficiently important blithely to arrogate to itself alone the fine title of man, which has perhaps been some what overused of late – here we find man suddenly feeling himself a stranger to his own science. He no longer recognizes his child, which has become too big, too powerful

---

[1]Lecture delivered September 5, 1952.

for him. He no longer knows what to do with the power that science puts in his hands, and he is even afraid he will misuse it. Man reproaches man for heaping upon him too great riches. He blames him for leading him too fast into the exploration of a universe which is too large and which at the same time has become too accessible, too docile. He is overwhelmed with too many riches, which has the effect of tempting him to squander them or even to destroy them. In the face of his own success, in the face of his own machines which have become magic slaves, man is afraid, and fear is a poor adviser.

Yet, my dear friends, what does this mean? Is it not a very humiliating situation for this man who is on the one hand a powerful sorcerer and on the other an apprentice intimidated by his own knowledge and terrified by the effect of his slightest gesture? Is it not time to re-establish a harmony, a unity, in this man's divided soul? For this, must we not lay the foundation of a new humanism, a humanism that would be total – including science – and would take the place of the classical humanism which was also total in its time?

I am rather tempted to repeat here, once again, an anecdote that is quite appropriate at the beginning of a communication such as this. It is the well-known story that describes Confucius being questioned by his emperor in connection with an alarming social crisis. Confucius is supposed to have answered, in substance, "In case of crisis, we must reform the nomenclature." We must do the same in the crisis that we are witnessing. It is essential to revise the names, in other words to undertake a precise definition of classical humanism or what it represents, and on the other hand, of science.

I do not conceal from myself the fact that this is a task that is far beyond me. I shall accordingly limit myself to proposing to you a method, or rather an approach to a possible definition of these two terms. I believe first of all that as it is usually formulated, that is to say statically, the problem is insoluble. More precisely, the data of the problem are then unintelligible; their dynamic aspect must be brought in. In other words, I believe that a crisis of our time cannot be judged if it is not first placed in the perspective of the evolution that has led to it, an evolution considered in its broadest sense, that is, taking in the evolution of all organized beings who have preceded man since life appeared on the earth.

Do not think that I am trying to minimize the problem of which I have to speak to you by comparing it with the immense sweep of the organic evolution of the world's history. No , no! It is, on the contrary, because I believe that the present situation, that is, the situation created by the appearance of science, is as new as that which occurred at the time when life appeared or when articulate language appeared among the animals. I believe that these are evolutionary turning points, not discontinuities but changes in the slope of the curve.

Let us then examine the principles that have presided over the evolution of organized beings, the first that has occurred on our globe. Three different principles can be distinguished, characterized by the words maintenance, multiplication and variation. I see that there are eminent geneticists in the front row; I hope they will not contradict me in the discussion that will follow.

The organized being, in fact, presents a structure that is different from that of inanimate natural objects, and it can be shown how the maintenance of such an organization, exceptional in its complexity, can be successful only through a constant struggle against the forces of destruction that tend to introduce disorder. I should like to quote in this connection the remark of a great physiologist, who defined life as "the combination of functions that resist death," that is, that maintain a certain complex state of organization. All living evolution, moreover, is based on a principle of identical reproduction. There can be no evolution without a multiplication of the individuals at each of its stages. The introduction of certain more or less important variations in the course of this multiplication makes it possible to pass from the static plan of individual maintenance to the dynamic plan of an evolution.

Let us consider for a moment the concept of maintenance, which is perhaps the most important. Such maintenance is conditioned by a certain degree of adaptation. There is no maintenance without "adaptation" to the environment in which the being is placed – and we shall take this term in a very broad sense. We shall say, for example, that a molecule that remains stable in a gas or in a solution, without changing its state, is adapted to the medium in which it is placed. If the environment changes in temperature or in composition and the molecule ceases to be adapted to it, the molecule will be profoundly trans formed or even destroyed. It is clear that the need to safeguard this relationship between the inner structure of the being and the environment will give rise to a selection among the variations, and we thus come back to Darwinism. Those among the new structures which are sufficiently well adapted to the environment maintain themselves, giving rise to natural selection.

In the course of identical reproduction, it is indispensable for the combination of the internal structures that are in a state of adaptation, that is, in correspondence with the environment, to recur in the descendants. While organic chemistry affords an elementary image of organic reproduction in the phenomenon of the autocatalysis of a substance by itself – a phenomenon in which a molecule causes the appearance of other identical molecules in an appropriate medium – it is not possible to conceive in so simple a fashion the reproduction of a large being whose adaptive structures involve a very large number of molecules simultaneously. A new stage appears between that of identical reproduction, confined to the molecular field, and that of the large living being, adapted to an environment. This stage is that of development, which makes it possible to pass from the details of a molecular structure (that of the chromosomes of the egg) to a macroscopic structure suited to be the seat of the physiological functions. This passage from the molecular to the anatomical occurs through a series of translations and amplifications. It remains true none the less that through the translation of development – a development that translates the molecular chemical properties of the genes and the chromosomes into terms of physiology and anatomy – a system of correspondence exists between the structural details of the initial chemical molecules and the external environment in which the living being must develop. There is a correspondence between the chemical functions of the molecule or rather of the group of molecules that form the heritage— the "germen" of living beings— and the outer environment, a correspondence that makes itself felt only by virtue of development and for the physiological and anatomical functions that appear with its help. There is a translation of the chemistry of the germinative molecule into physiological terms, a translation that makes adaptation to the environment possible.

It is perhaps because of this necessary translation that evolution has been very slow. Every time that a molecular chemical modification appeared in the germinative part, it was necessary to wait for developments to occur so that the translation might be appreciable, in other words to wait for the variations on a molecular scale to appear in the form of new anatomical, physiological structures of the developed being. It is thus certain that the mechanism of evolution must be greatly accelerated when this stage is not necessary, that is, when identical reproduction oc curs directly for the whole being, as is the case for a certain number of beings that I shall venture to call living, even

81

though Professor Guyenot believes that they are not yet altogether so, namely the viruses. These are in reality large organic molecules that reproduce themselves directly in a suitable environment without for this reason constituting a more considerable living being. A variation in the chemistry of one of these molecules immediately makes itself felt in its adaption: after the variation it is directly more or less adapted to the environment, and as a consequence it disappears or multiplies. It is clear that evolution proceeds more rapidly in this case.

If it is true, as has sometimes been suggested and as I myself have proposed in a recent book, that the evolution of ideas merely follows and extends the evolution of living beings, and if we may resolutely take the point of view of the geneticist in studying the evolution of these ideas, we may perhaps attempt to base on genetic considerations of this nature a distinction between several categories of ideas – I am tempted to say: between different strains of ideas, since, in the present theory, they form strains which, like strains of animals, multiply by passing from one brain to another. They people a population of brains with a population of ideas. Let us leave aside the specifically creative mechanism and concentrate on the device of selection, which is the one that interests us this evening and which involves genetics. It is the one which, among the various novelties that have appeared in the course of the work of the mind (or idealization), makes a choice that binds the future. This choice must take into account the imperious necessity to reproduce ideas in numerous copies in the brains of men, by virtue of articulate language; it is clear that an idea that is not transmissible, that does not multiply by passing from one brain to another, does not belong to one of these strains, however useful it may be. Even if it is an excellent idea, if it is not transmissible it dies with its creator.

When we speak of selection, we must base it on a precise consideration of the criteria that define it. These are generally criteria of utility: thus Darwinism is based on the utility of the newly appearing characteristics. But they are also at times criteria of satisfaction of a more subtle order; we shall later see certain applications. The type of selection will differ considerably depending on whether the utility in question is relative to the individual who has the new idea (or who transmits it), or relative to the groups that several individuals of the same species form among themselves, or relative to the entire species. In other words, and since we have characterized adaptation as a precise correspondence between an internal structure – which was a chemical structure in the case of the molecule, a physical, physiological and anatomical structure in the case of living beings, here an idea structure – and external conditions, the adaptation must occur either directly between the ideas of an individual and the conditions of the environment that surrounds him, or in directly, through a combination of individuals whose ideas constitute a traditional inheritance. These ideas that are peculiar to the group share its fate, whether it prospers or deteriorates.

Inasmuch as the problems of existence of a group are very often problems of struggle against other groups of a similar nature, it is natural that the general body of ideas selected to form a part of the tradition of a group should above all strengthen and consolidate the group itself. In many cases the individual will be deliberately sacrificed for the maintenance of the group. This is what happened in the course of the selection of instincts among the social insects, to say nothing of men, with the difference that in the former case the selection related to hereditary characteristics and in every instance had to wait for the complete development of the society (ant hill or hive) before making its indirect action felt. Many hereditary characteristics of the individual, ant or bee, are harmful to the individual as such, since he must at times sacrifice himself to defend his society. This is an inner disposition that is disastrous for the individual, but necessary for the maintenance of the group.

The same thing is true in the case of ideas that have been selected in the light of such collective criteria. These ideas are favorable to the maintenance of the group, but may well be disastrous, unfavorable, fatal, harmful for the individual as such. Among the ideas that will result from a selection of this kind must be included ideas of the moral type. They will include myths, tales belonging to the traditional folklore, in short everything that involves the group as such. To these must be added a technological tradition, which enables the group to attain a certain living standard for its members.

But the point I should particularly like to emphasize is the following: In the natural selection of living beings, it may be said that every viable being survives: biologists know many examples of living beings which appear to be rather poorly adapted, like certain birds which have an excessively large beak or extraordinary ornamentation; paleontology shows us the existence of large saurians bearing scales whose great weight must have hindered them considerably in their movements. The study of evolution brings to light numerous examples of species which, as a result of a certain number of variations, end up by becoming less and less well adapted, but which nevertheless continue to live and to reproduce so long as they are viable. There comes a time, however, when the species disappears.

In the same way it may be said that every traditional idea that is easily transmitted, that is not harmful to the group in which it is implanted, maintains itself: it is viable. There is an ad minimum selection, which suppresses the really harmful ideas or combinations of ideas, but only these, and which allows all the ideas that are above a certain adaptation minimum to subsist. Just as a characteristic that has proved baneful in an individual of a living species and has caused this individual to disappear may very well subsist in a neighbor which does not find itself exactly in the same conditions, so an idea considered false by one group of men may subsist in a neighboring group when it is an idea that has been selected by the group.

In the category of scientific ideas we shall find characteristics radically opposed to those that we have just described. Going back to the genetic analysis, we may say that scientific ideas behave like a strain whose members are exactly alike because they stem from a single pair of ancestors. For such a lineage to be possible, it is necessary for each of the ideas, each of the concepts, to be quite strictly definable: the transmission of this idea by articulate language will always give an identical idea in each of those who carry it. We know that men and groups of men differ profoundly; it is therefore not easy to find standards that can serve as bases for these definitions. But when we are dealing with scientific ideas, it is possible, thanks to the device of precise correspondence between internal structures and external facts. This exact correspondence, which makes it possible to establish the rigor of scientific ideas, thus endows them with an objective character. They then become universal; that is, they become the attribute of the entire human species and not only of a group of individuals.

But every medal has its reverse side and the universality of scientific ideas, due to their rigor and their constant comparison with the external environment, entails on their part an extreme sensitivity to numerous causes of death. To revert to genetic terminology again, when we are dealing with a population all of whose elements are identical and external conditions change in such a way that one representative of this population dies, all die. Likewise, when a scientific concept is no longer tenable because the correspondence with certain characteristics of the external universe on which it was founded is demonstrated to be false by a single man, this concept is rejected by him and must be rejected by all other men as soon as the proof of its incorrectness has been brought to their knowledge. It is, in short, death that is multiplied, or rather it is a new idea contradictory to the first that multi plies and that takes its place. The rigor and objectivity of scientific ideas thus consecrate both their universal value and their susceptibility to contrary proof. How different from nonscientific ideas and concepts! These, which are the attribute of groups, have a robust vitality, which they owe essentially to their subjectivity and also to their lack of preciseness: contrary proof is hardly applicable to them. Moreover, even when a man is convinced of the incorrectness of one of these ideas, he is incapable of transmitting this assurance to his fellows with finality.

Scientists and their more or less abstract edifices have often been made fun of, because from year to year (and sometimes from day to day) certain of their hypotheses were seen to collapse, only to be replaced by ideas that were found to be better. How can we have confidence in this or that new theory, since we see it blithely taking the place of the preceding theory, now abandoned? This is the price of objectivity and universality. Conversely, we see ideas, such as those at the basis of spiritism, telepathy, astrology, resist all the assaults of science and numerous administrations of contrary proof. They are simply not susceptible to these by virtue of their device of selection. There are, even now, people who come into the physics laboratories speaking of N rays. Yet this has to do with results arrived at by men of science, which were subsequently proved to be invalid; for scientists such occurrences are normal and leave no doubts; these ideas have simply been abandoned. But the nonscientist is not sensitive in the same way to contrary proof and may continue undisturbed to speak of N rays or to expatiate on mitogenetic rays. On the other hand, you will never see a man of science affirm that the Ptolemaic system provides the true mechanism of the solar system.

How does it comes about that man, in his effort at adaptation (that is, I repeat, in his effort to establish a system of precise correspondence between his ideas de rived from his internal structure and the phenomena of nature that surround him), has not from the beginning applied a system of rigorous selection that would have made it possible to build a scientific system at the outset?

It is not that primitive man, the man of prehistoric times, did not establish certain correspondences that were precise, universal and consequently scientific. For example, those that enabled him to develop precise techniques (tool making, devices for hunting and fishing, agriculture). But these precise correspondences remained isolated, sporadic, each valid in itself. They do not form a general system, so that the internal counterparts of these natural phenomena (ideas) were not linked to one another. They were transmitted isolatedly, as such, from generation to generation, like wholly separate animal species.

The reason for this lack of inner links is easily explained by the extraordinary complexity of the intellectual device that later proved necessary to effect them. The problem was not to link the ideas among themselves by means of some internal mechanism. The latter must correspond *as a mechanism* to the external mechanism that connects the phenomena in question to another. An exact parallel must be established between the two.

This liaison was made possible only after a reflective study of his own internal structure by man himself. This study enabled him to discover what were the types of link that he himself could establish among different ideas and to seek among these devices – sanctioned by his internal structure – those that constitute a parallel with the links existing in nature among corresponding external phenomena.

This long and difficult study is that of logic and mathematics. Before it had gone very far and in order to satisfy the craving for a system, man sacrificed the precision and the objectivity of correspondences to the establishment of any kind of links among his ideas so long as they satisfied him. The myths, mysteries, cosmogonic theories of antiquity could satisfy this need of adaptation at little cost, and on the other hand they could victoriously resist even repeated failures, since they are not within the category of scientific ideas. It was only through the intermediary of groups and the selection of these groups that these systems could finally be somewhat selected.

Confronted by these facts, must we not at once ask ourselves the contrary question, in other words, wonder why man has not always been content to have recourse to empirical relationships, connected by some device or other, why he has not been content to satisfy his need of communion with the world by means of imprecise inner systems, vaguely relating a certain number of ideas in correspondence with external phenomena? This state of humanity did in fact last from the appearance of the first communities up to the Renaissance, and it really changed only with the advent of the scientific era.

There was first a development of empirical techniques which continually increased the number of isolated correspondences between ideas and external phenomena. A n d this increase in number made the question of the relation of the corresponding ideas among themselves, of their intelligible relations, ever more pressing and led to the development of the precise study of its own internal structure by the human mind. Thus it was the development of these empirical correspondences on the one hand and the progress of the study of man by himself on the other that suddenly brought about the concept of this parallelism between the grouping of internal ideas and the grouping of external phenomena. The comparison of causal chains among external phenomena with the inner logical chains of ideas that correspond to them constitutes the scientific system properly speaking. It only remained to accept the price of this rigor and this precision, that is, the mortality of systems thus obtained, their fatal susceptibility to contrary proof.

But even granting this whole mechanism that I have just described, there still arise many fundamental questions. Thus, should we not wonder whence this thirst for explanations, that is, this thirst for adaptation, springs? And it is a thirst that can at times be quenched in such a trivial way, through a few artifices of language or vague, distant analogies!

I should like, in this connection, to relate a personal anecdote that made a deep impression on me. After I had delivered two lectures at the Mediterranean University Center in Nice on the subject of cosmic rays (needless to say, the audience was much smaller at the second session than at the first), one listener came up to me at the end and said, "I should like to tell you what I think of life." He then explained to me with a great deal of passion that he visualized life as "whirls of cosmic rays." That was why he had come to talk to me. As I seemed a little surprised, he said eagerly, "I don't expect to convince you, but I like this theory; I explain life by whirls of cosmic rays, and this satisfies me." Then he hurried off, lest I attempt to demonstrate to him the inanity of his theory. He wanted to talk to me in order to have the satisfaction of once again expressing his theory; then he hastened to shut the lid on his treasure and to make off with it. It is strange that the adaptation of the microcosm and of the macrocosm should be such that a man can content himself with three or four magic words, taken almost at random, with a few high-sounding terms, by means of which he builds his inner life.

But it is not our concern this evening to seek why this

thirst for knowledge and adaptation exists, and I should like to turn toward another very important question, contained in the simple observation that science is possible. It is possible to find inner chains of ideas adequate to the structure of our thought, since they form a satisfying inner system, and since they correspond precisely to sequences of phenomena observed externally. Conceivably this might have been impossible. Must we not see in this observation the proof of the existence of the external world? And even, because of the similarity of structure between this world and our thinking, a proof that the latter belongs to the external world?

If we can find sequences of subjectively satisfying internal structures which, on the other hand, are parallel to the objective sequences of external phenomena, is this not a sound foundation for a kind of monism – which would, however, differ greatly from a too narrow materialism – and which would perhaps save us from the threat of intellectual schizophrenia that the absolute separation between the world of things and that of the spirit constantly holds over us?

Having thus established science in the great current of the evolution of beings, perhaps even at the peak of their increasingly complex and precise adaptation to the phenomena that surround them, let us look into the question of the limits of its realm: Whither can the evolution of scientific ideas lead? Is it limited in its advance, and if so by what principles? On the other hand, scientific ideas constantly bordering on other types of ideas and inter actions are inevitable. Where, then, are the frontiers?

The first problem, that of the advance of evolution and its possible limitation, in reality presents two quite distinct aspects. It may be wondered – and the question has arisen very insistently – whether science can continue to progress at the present increasingly rapid rate. Will it not be obliged to slow its advance or perhaps even to stop it, when all the phenomena accessible to man are linked to one another by satisfying theories? On the other hand, considering that the complexity of our ideas depends directly on the structure and the mode of functioning of our brain, we may wonder if, in the course of building increasingly complex, increasingly refined structures, we shall not reach a limit imposed by this internal structure itself.

As for the possible exhaustion of the resources of nature, I believe there is no anxiety to be felt on this score – or no hope, depending on the point of view. There is in any case not the slightest indication in this direction. New realms of science open at every moment in physics, in biology, in chemistry, to say nothing of the social sciences, which have barely been outlined. Yet science has itself discovered a number of the natural limits that restrict the field accessible to man.

Astrophysics, for example, has revealed that part of the universe is forever inaccessible to us, whatever the means used to seek knowledge of it. I am speaking of that portion of it that is situated very far from us and which, because of the phenomenon of expansion, is vanishing very rapidly. The observation of the stars and nebulae that are progressively more remote from the earth shows, by the dis place-ment of the spectrum of their light, that they are moving away from us more and more rapidly and proportionally to the distance. There is thus a distance beyond which the celestial bodies move away from us so fast that their light can no longer reach us, which does not contradict the principle of relativity.[2] We can therefore receive from

---

[2]EDITOR'S NOTE: The wording of this sentence may be misleading despite the clarification at its end. Readers without proper physics education may interpret the phrase – "the celestial bodies move away from us so fast that their light can no longer reach us" – in terms of everyday language – that it is the enormous velocity of the celestial bodies that prevents their light from reaching us. Those bodies' receding speeds are *apparent* as a result of "the phenomenon of expansion" as Pierre Auger points out above.

them no light message and we can send them none. This is really part of the universe which we know exists but which is completely inaccessible. Certain nebulae are at the very border, are about to drop below that new type of horizon; they are the ones that are moving away from us so swiftly that their light has already assumed a red tinge, and the physicist knows, merely by looking at the spectrum of this light, that they will soon have disappeared forever. Soon – that is to say in a few million years.

After this practical limitation of the knowledge of the universe, let us give a more abstract example. Satisfying theories have sometimes been proposed to unite and ex plain many varied phenomena. It unfortunately happens that the verification of certain theories would require conditions that we cannot fulfill, such as very precise observations over a very long period, for example a billion years. This also is beyond our scope, although I do not mean to imply by this that in a billion years there will no longer be men on earth. These theories are perhaps correct, but we cannot allow ourselves to judge them with finality.

Apart from a few special cases of this kind, all the prophets who have successively attempted to exclude some realm of nature from science have always been gain said by facts. I shall give only one example, that of chemistry. Berthelot had been told that, despite the impetus that he had given to chemistry, he would not be able to manufacture organic substances, because this was the prerogative of life. Nevertheless, he soon synthesized acetylene and later alcohol. Then the supposed barrier was displaced, and it was sugar that was put beyond the bounds of chemistry. Yet Fischer synthesized sugars. Will we have to stop at the proteins? No, for chemists are linking the amino acids to one another, and before long it will be possible to manufacture this or that protein at will.

There are, *a priori*, no realms "forbidden" to science.

What is much more probable is that science will become increasingly less accessible, in its entirety, to a single man, and that it will become necessary to invent more and more powerful synthetic methods of presentation in order to make general views accessible, at least to a few. The scientist of the future will no longer dream of mastering the whole of science. He will not need to, if the theoretical constructions are broad enough to reduce all the essentials of science to accessible syntheses.

The question of structure is much more serious, and perhaps we have already reached, in mathematics, the end of our rope. It is our faculty for representation that is first exceeded, without depriving us of the possibility of reasoning. We can, for example, define and study hyper spaces to n dimensions that cannot be represented concretely. I wonder, in fact, if we do not have an example of this difficulty with the particle-wave crisis – as it has been called. It is not impossible that the whole of these phenomena that we want to explain cannot be integrated into a single system by our internal structure; that the latter allows us to integrate only a single aspect of these, through a system of correspondences – that of corpuscles – and then another, through another system – that of waves. We are perhaps unable to integrate the whole with a single internal combination. If we are limited in this direction, this does not at all mean that we are thereby prevented from reasoning correctly, but that we cannot reason on all the aspects at one time, with a single system of correspondence: we need several to reconstitute the total system.

Though I said a moment ago that no realm is forbidden to science, this does not mean that science will resolve all problems and will answer all questions. Very often problems are posed in contradictory terms; they have no solution because they have no meaning. Or they violate the laws of nature in their very premises. Often, too, the answer will leave us wholly unsatisfied, as when science answers us with a probability when we were expecting a certainty. If it does not satisfy us, it nevertheless gives us an answer that is the only possible one.

Having examined the possible limits of science, I now come to the study of its frontier relations with other forms of activity of the human mind. And first of all, the impact on the external world. Whereas animals are con tent to adapt their bodies as best they can to the conditions in which they are placed, man continues this adaptation. He does so on the one hand by setting up a system of ideas in correspondence with all aspects of the universe and on the other by modifying the environment that surrounds him, in order to bring it into greater harmony with his own inner structure. Animals also act on their surroundings to make them more hospitable, but they always act in accordance with the same plan fixed for each species by heredity and by instincts. Here too, man eludes organic evolution and contributes new elements through individual creation, and he does so along two lines of evolution of quite different significance.

The one corresponds to artistic production; through it certain aspects of the external world can be given a form that proves to correspond directly with certain internal elements that are a part of our affective heritage, such as memories, associations having a sentimental value, or even with vital needs like hunger or sleep. The artist, by virtue of his constructions, transforms the world that surrounds him – a world which, without being deliberately hostile, is normally indifferent if not unfriendly – into a hospitable world in which we recognize at every moment the projection of our own internal structure.

We sometimes see landscapes that may be very beautiful, but with a trace of anguish, because of their wildness. Such landscapes can be transformed by a Le Nôtre into parks. He lays out roads, makes the thickets bloom, isolates large trees on lawns, places statues and benches inviting repose. The landscape has become a universe humanized both in its details and in its main lines. This is the role or art.

The other direction in which the world can be changed is that of technology. The correspondence between objects manufactured by the technologist and man's internal structure is no longer a direct correspondence this time, for it passes through an activity of his body. The machine enables man to do better, faster, with less expenditure of energy, what he might have undertaken with only the resources of his body. And this develops more and more through the degrees of the hierarchy of the machines, from the simplest tool to the great instrument for rapid calculation or long-distance transport.

Leaving aside the question of the relation between art and technology, let us turn toward the problem of the relations between the sciences and the teachings of morality. The manner in which this question is usually treated is essentially negative. Science, we are told, is not normative, it cannot give rise to any categorical imperative, and the privilege of guiding man in his decisions is reserved solely to considerations of moral or religious values. Science, at most, is accorded the right to help man in his actions with its power, so that he may be better able to apply with full knowledge the imperative rule that he must follow. Most representatives of science blithely adopt this attitude, how ever negative, because it thus enables them to evade a very grave problem with which they are deeply concerned.

Recently there has developed among the general public a tendency to consider particularly the harmful aspects of the applications of scientific technology and to attribute to them a kind of infernal virtue. It is science as such that is supposed to be responsible for the destructions of the world wars; it is science that has introduced those frightful novelties in armament known as asphyxiating gases and the atomic bomb. The scientists who have attempted to defend their science against such accusations have been content to fall back on the generally accepted opinion that deprives science of all access to moral values, leaving to it no other role than that of increasing the power of man over nature. By withdrawing in this way, by refusing contact, they can avoid answering the questions that they are asked. A n armed man, they say, is always dangerous. Is there a very great difference between an entire city put to the sword, pillaged, burned by the soldiery as was done one or two centuries ago, and a few thousand lives destroyed by an atomic bomb, suddenly, during an aerial bombing? What really counts is the ideologies that are at the basis of human action: cruelty, violent feelings. The instrument used by the man who wants to do harm is of only secondary importance. The two realms have been separated and everyone appears happy.

Yet this is merely a defensive position, a negative position. Is it not possible to go further in the search for a synthesis between the elements that determine men's actions? We must not be satisfied with a defeat of this kind that would sanction what can be called schizophrenia, dividing our brain as it does into two parts. I propose that this synthesis, which would bring together the two elements, be achieved, not by attributing to science a normative role, that is, by extending the realm of science toward that of ethics, but on the contrary by bringing the role of the latter toward informative actions, that is, toward science. This point of view is in fact not new, but it has perhaps not been considered rigorously enough to be truly effective.

Essentially, the problem involves free will. If man, for each of his actions, can make a choice, he will do so in the light of a certain number of elements of information. Among these will be elements of rational, scientific information whose role as such no one will dispute. But there will also be moral elements. Thus the fear of the reprobation of others or the desire to be admired. This does not mean that man will act only under the sway of the fear of hell, of prison or of the contempt of his fellows.

Very often the coercive power of this information will have been forgotten because it has been lost in a too distant past. It will have left only a sort of conditioned reflex. We will act under the influence of an impulse that we do not rationalize and that we feel as a categorical imperative, but which is in reality an earlier information, whether it has come to us in the course of our education or been given to our teachers who have then transmitted this imperative to us. These are, to all intents and purposes, conditioned reflexes that have become traditional automatisms. This being the case, is it really possible to draw a well-defined line between scientific information and moral information which together define the atmosphere within which a man must make his decisions? It appears to me that we might derive from the principles set forth at the beginning of this lecture, which constitute a kind of genetics of ideas, a criterion of distinction, on condition that we consider not only the present state but the evolution that has preceded it.

We have seen that scientific ideas were selected in such a way as to make possible a precise, universal correspondence with natural phenomena, depending neither on the group of which man is a part nor on his traditional training, but only on the nature of this man and on the structure of his brain. On the contrary, ideas of the moral type have been selected through the group or groups of which the person considered is a part or of which his ancestors or his teachers have been a part. This amounts to saying that these ideas do not offer a precise, universal correspondence with external phenomena, but that it is their repercussions on the behavior of groups that are in correspondence with the universe and that adapt these groups to the environment in which they are placed. I should not want to seem to solve by means of a few sophisms the question of the conflict between ethics and science; accordingly I offer you only some simple indications of method. It seems to me that the attempt to bring science and ethics more closely together, that is, to build normative precepts and rules on scientific knowledge, is the wrong approach to the problem.

What I propose, quite on the contrary, is to bring ethics closer to science by reinvesting it, by means of the genetic theory of moral ideas, with its informative character, present or past. If we accept the disappearance of categorical imperatives, if we admit that man always has a certain margin of freedom and that his definitive choice is based on a more or less disguised information that sometimes has passed to the stage of a reflex, the problem will be, if not solved, at least stated on bases that do not rule out the hope of building an acceptable solution in the future.

But I realize that I am perhaps taking advantage of the signal privilege conferred by the position of lecturer, the privilege of being able to speak without being interrupted or contradicted, without being thrown off the track by interruptions, expected or unexpected. Many of you must be thinking at this moment: well and good, as far as information is concerned, but what about values? It is through these that the normative quality of ethics makes itself felt. It is these that guide man's decision, after all the information has been given and he has weighed the arguments for and against. Your analysis sidesteps the main issue.

We shall surely debate this problem in the course of the discussions that are to follow, but I should merely like to defend myself against the sin of omission. I have not forgotten values, even though I have not yet spoken of them. What I have to say on this score may appear to many so scandalous that I am not without anxiety at the moment of formulating it before you. In order to lessen this great scandal, I should like to prepare you for it by a small scandal whose terms are, in fact, not my own: very often, philosophy is nothing more than fossil science. It is ideas of a scientific kind which have lost that quality be cause they have been accepted too long without discussion and which finally are no longer questioned. They have become respectable and impressive, like old men, but old men from whom one no longer receives or expects any new teaching. This is the first scandal; and now, like Flamineo, I am damned and I can therefore allow myself whatever I wish. Let us go on to values.

Values, generally speaking, are fossil information. They are ideas of the moral type and even ideas of the rational type that have aged in ourselves since our youth and also aged in our group since their introduction. They have become respectable, unchallengeable, sacred. A s their origin has been lost, they appear at first as transcendental.

A comparison with psychoanalysis may be helpful here. A patient comes to be psychoanalyzed: He has uncontrollable impulses that transcend his will and against which he is powerless. They are *a priori* imperatives that he cannot resist. The doctor explains to him the psycho analytic method and discovers the origin of these impulses in a past, distant or relatively recent, but in any case forgotten. He brings out and holds up to the light, as it were, this fossilized item of information, whose origin the patient himself has forgotten. As soon as the patient has understood that this *a priori* impulse, which he experienced as something magic and which obliged him to act, was nothing but an old item of information resulting from events that he has forgotten, he is cured.

Perhaps we need to cure ourselves of certain values by clearly recognizing their origin. We would, it may be added, replace them at once by others, which our present information would lead us to create in order to guide us in our actions. It our descendants forget these items of knowledge, these present reasons that impelled us to choose this or that way, and accept the injunctions with out being aware of where they come from, they will have reconstructed incomprehensible values, which they will obey out of respect for the sacred.

Let us take an example. Doctors have discovered processes of anesthesia that render many operations painless. They have then attempted to apply this method to child birth, and as the results at the outset were not perfect, there were people who accused them of destroying a value, because woman is supposed to bring forth children in sorrow – or pain. This pain, having in this special case a particular value, must not be decreased artificially. Medicine has nevertheless progressed and has discovered harm less ways, both for the woman and for the child, making more satisfactory childbirth possible. Consequently, and fortunately for mothers, this tabu has been relinquished almost everywhere.

An objection often made to developing scientific information at the expense of a realm that until now was re served to considerations of another order resides in the extreme complexity of the situations in which a man may be placed because he belongs to one or several groups.

When it is a question of measuring a physical magnitude, a suitable instrument on which everyone is agreed suffices: one looks at an ammeter, one takes the reading. But if it becomes necessary to take into account the inter dependence of various personalities, the physical, mental capacities of various persons, to evaluate the consequences of an action throughout an administrative or family hierarchy, it is no longer possible to have recourse to a scientific method that will give the correct information, and the experiment is of no help, since what we have is a unique experiment, a unique situation.

Without venturing to solve or to indicate ways of solv ing these questions, I should like to bring to your atten tion the example of a case in which it was recently demon strated that if the necessary effort is made, scientific in formation can assume a much more considerable place than it had been vouchsafed up to the present; I am referring to the introduction of operational research among the elements of decision of the military command.

Numerous army chiefs have had recourse to technologists and scientists in order to inform themselves as to the conditions under which a defensive or offensive action may be undertaken. In operational research there must be a thorough recasting of all scientific or parascientific ideas scattered among military theories, in order to introduce precise quantitative elements into the information presented to the command.

First of all, I wish to point out that the conclusions of the operational investigation are indeed presented as information, since they are contained in a report delivered to the desk of the commander-in-chief. Afterward the scientists who have done the research leave the command to decide the action to be undertaken on the basis of the information given. There is no categorical imperative in science. I recall in the second place that the conclusions of the operational investigation are delivered to the command only after a thorough survey, during which the scientists endeavor to introduce *all* the elements, without exception, that may be instrumental in ensuring the success of the undertakings.

The scientist must break through all the hierarchical barriers, for he places on the same level the answers of the enlisted man, of the general or even of the civilian... No file, no establishment, no advice must be closed to him, and it is only from a synthesis of all the significant elements that he will have brought together that the investigator will at last be able to draw a conclusion that will have the required soundness and objectivity. It is worth noting that the scientists who have been most successful in this difficult task are not necessarily those who deal with the most abstract sciences. It is not the mathematicians but rather the biologists who are perhaps the most effective. This is because they are more accustomed to taking into consideration a very large number of parameters of unequal value. In several cases it is they who have given the correct solution, far removed though it may be from the biological field. It was a scientist who indicated the dimension of the convoys to be formed with the escorted ships when they crossed the Atlantic; it was neither a seaman nor a soldier. The success was such that the method was extended to quite different activities, like industrial production, for example. Here too the problem was not exactly to wipe out all tradition but rather to bring in these traditional elements, with their psychological value, on the same footing as purely scientific elements, that is to say to consider them objectively, to admit their variable nature, to propose trans formations of them.

It is possible, alas, that even as operational research has been considered to be purely scandalous in its methods and in its conclusions by numerous military men, the introduction of the scientific method and spirit into the *totality* of elements entering into a human situation may be intolerable to many. Yet must we let watertight partitions, like those I referred to a moment ago, be set up be tween scientific information and nonscientific information without attempting, through an effort at unification, to establish possibilities of comparison and a reciprocal influence between the different types of information?

The thing appears not so very difficult insofar as the relations between information of the scientific type and that which derives from the physiological functions of our organs, that is, data perceptible to the senses, are con cerned. A study of innate or conditioned reflexes, research on the influence of internal secretions on the activity of the mind, already constitute a part of the scientific pro gram of many laboratories and institutes. We may hope in this way to obtain a precise and scientific knowledge of the whole of these influences and to deal with them objectively. The question arises as to how to overcome certain natural items of information by scientific items of information, which we do for example when we are given a very bitter medicine.

The natural information indicates, "It is not good" and makes us spit it out, while scientific information tells us, "You must swallow it because it will cure you..." We can imagine that in this way there might be created a perfectly reasoned inhibitory device that might be very effective. It is probable that study would also lead to the recognition of the necessity of letting certain actions occur freely in the light of information of the natural type, of the physical type, in order to maintain adequate function. This is what has often been described as "safety valves."

We have, moreover, a well-known precedent in the adventure of Ulysses with the sirens. Rational, scientific information led him to deafen all the crew of his ship. Since sensual information tempted him to listen nevertheless and his intelligence enabled him to foresee his own irrational impulses, he placed himself under conditions that were such that no harm could result. Like Ulysses, we place ourselves from time to time in a situation in which we cannot do anything foolish before removing the wax from our ears and listening to the sirens. We renew a passionate contact with the savagery of our primitive nature, but if we are well tied, nothing irremediable happens.

The problem is a good deal deeper when it involves the relations between scientific information and that which is contained in the tradition of groups and which is at the root of their social functioning. It is no longer sufficient to plug one's ears and to develop inhibitions in order to be able to resist the urgings of one of the forms of information, on condition that one give oneself respite from time to time.

Ordinary life presents us at every moment with contradictions between the counsels of scientific reason and the directives of tradition. The rational manner would consist first of all in bringing down numerous values from their transcendental pedestal to confront them on an equal footing with other scientific or sense "parameters." The reasonable choice would then often appear quite clearly. Y e t in the last analysis does there not remain at least one very general value that carries the decision, at times unbeknown to us?

I myself have proposed an ethics of creation that would keep man in the great strain of evolution toward the reciprocal adaptation of the microcosm and the macrocosm, through scientific ideas, inventions, the arts. But one may not like this rather dizzy evolution and prefer the stable and no less real values of maintenance. Is it not, in the last analysis, temperament that determines the direction? And does this temperament not bring us back to genetics and to physiology?

But whatever may be the answer to all these questions, I should like at least to have shown you once again what a fine task may be assigned to a total humanism of the future: that of giving back to man at last a unity that he had undoubtedly known only once, at the very beginning of his evolution, when his physiological reflexes acted alone. At that time he had no anxiety because he obeyed only his instincts. It was perhaps a golden age, when man did not yet formulate problems!

If a golden age existed in the distant past, it may also be envisaged for the future, but this time it would be based not on elementary needs, but on thought.

THE DEVELOPMENT OF QUANTUM MECHANICS

NOBEL LECTURE, DECEMBER 11, 1933

WERNER HEISENBERG

Quantum mechanics, on which I am to speak here, arose, in its formal content, from the endeavour to expand Bohr's principle of correspondence to a complete mathematical scheme by refining his assertions. The physically new viewpoints that distinguish quantum mechanics from classical physics were prepared by the researches of various investigators engaged in analysing the difficulties posed in Bohr's theory of atomic structure and in the radiation theory of light.

In 1900, through studying the law of black-body radiation which he had discovered, Planck had detected in optical phenomena a discontinuous phenomenon totally unknown to classical physics which, a few years later, was most precisely expressed in Einstein's hypothesis of light quanta. The impossibility of harmonizing the Maxwellian theory with the pronouncedly visual concepts expressed in the hypothesis of light quanta subsequently compelled research workers to the conclusion that radiation phenomena can only be understood by largely renouncing their immediate visualization. The fact, already found by Planck and used by Einstein, Debye, and others, that the element of discontinuity detected in radiation phenomena also plays

105

an important part in material processes, was expressed systematically in Bohr's basic postulates of the quantum theory which, together with the Bohr-Sommerfeld quantum conditions of atomic structure, led to a qualitative interpretation of the chemical and optical properties of atoms. The acceptance of these basic postulates of the quantum theory contrasted uncompromisingly with the application of classical mechanics to atomic systems, which, however, at least in its qualitative affirmations, appeared indispensable for understanding the properties of atoms. This circumstance was a fresh argument in support of the assumption that the natural phenomena in which Planck's constant plays an important part can be understood only by largely foregoing a visual description of them. Classical physics seemed the limiting case of visualization of a fundamentally unvisualizable microphysics, the more accurately realizable the more Planck's constant vanishes relative to the parameters of the system. This view of classical mechanics as a limiting case of quantum mechanics also gave rise to Bohr's principle of correspondence which, at least in qualitative terms, transferred a number of conclusions formulated in classical mechanics to quantum mechanics. In connection with the principle of correspondence there was also discussion whether the quantum-mechanical laws could in principle be of a statistical nature; the possibility became particularly apparent in Einstein's derivation of Planck's law of radiation. Finally, the analysis of the relation between radiation theory and atomic theory by Bohr, Kramers, and Slater resulted in the following scientific situation: According to the basic postulates of the quantum theory, an atomic system is capable of assuming discrete, stationary states, and therefore discrete energy values; in terms of the energy of the atom the emission and absorption of light by such a system occurs abruptly, in the form of impulses. On the other hand, the visualizable properties of the emitted radiation are described by a wave field, the frequency of which is associated with the difference in energy between the initial and final states of the atom by the relation

$$E_1 - E_2 = h\nu$$

To each stationary state of an atom corresponds a whole complex of para- meters which specify the probability of transition from this state to another. There is no direct relation between the radiation classically emitted by an orbiting electron and those parameters defining the probability of emission; nevertheless Bohr's principle of correspondence enables a specific term of the Fourier expansion of the classical path to be assigned to each transition of the atom, and the probability for the particular transition follows qualitatively similar laws as the intensity of those Fourier components. Although therefore in the researches carried out by Rutherford, Bohr, Sommerfeld and others, the comparison of the atom with a planetary system of electrons leads to a qualitative interpretation of the optical and chemical properties of atoms, nevertheless the fundamental dissimilarity between the atomic spectrum and the classical spectrum of an electron system imposes the need to relinquish the concept of an electron path and to forego a visual description of the atom.

The experiments necessary to define the electron-path concept also furnish an important aid in revising it. The most obvious answer to the question how the orbit of an electron in its path within the atom could be observed namely, will perhaps be to use a microscope of extreme resolving power. But since the specimen in this microscope would have to be illuminated with light having an extremely short wavelength, the first light quantum from the light source to reach the electron and pass into the observer's eye would eject the electron completely from its path in accordance with the laws of the Compton effect. Consequently only one point of the path would be observable experimentally at any one time.

In this situation, therefore, the obvious policy was to relinquish at first the concept of electron paths altogether, despite its substantiation by Wilson's experiments, and, as it were, to attempt subsequently how much of the electron-path concept can be carried over into quantum mechanics.

In the classical theory the specification of frequency, amplitude, and phase of all the light waves emitted by the atom would be fully equivalent to specifying its electron path. Since from the amplitude and phase of an emitted wave the coefficients of the appropriate term in the Fourier expansion of the electron path can be derived without ambiguity, the complete electron path therefore can be derived from a knowledge of all amplitudes and phases. Similarly, in quantum mechanics, too, the whole complex of amplitudes and phases of the radiation emitted by the atom can be regarded as a complete description of the atomic system, although its interpretation in the sense of an electron path inducing the radiation is impossible. In quantum mechanics, therefore, the place of the electron coordinates is taken by a complex of parameters corresponding to the Fourier coefficients of classical motion along a path. These, however, are no longer classified by the energy of state and the number of the corresponding harmonic vibration, but are in each case associated with two stationary states of the atom, and are a measure for the transition probability of the atom from one stationary state to another. A complex of coefficients of this type is comparable with a matrix such as occurs in linear algebra. In exactly the same way each parameter of classical mechanics, e.g. the momentum or the energy of the electrons, can then be assigned a corresponding matrix in quantum mechanics. To proceed from here beyond a mere description of the empirical state of affairs it was necessary to associate systematically the matrices assigned to the various parameters in the same way as the corresponding parameters in classical mechanics are associated by equations of motions. When, in the interest of achieving the closest possible correspondence between classical and quantum mechanics, the addition and multiplication of Fourier series were tentatively taken as the example for the addition and multiplication of the quantum-theory complexes, the product of two parameters represented by matrices appeared to be most naturally represented by the product matrix in the sense of linear algebra – an assumption already suggested by the formalism of the Kramers-Ladenburg dispersion theory.

It thus seemed consistent simply to adopt in quantum mechanics the equations of motion of classical physics, regarding them as a relation between the matrices representing the classical variables. The Bohr-Sommerfeld quantum conditions could also be re-interpreted in a relation between the matrices, and together with the equations of motion they were sufficient to define all matrices and hence the experimentally observable properties of the atom.

Born, Jordan, and Dirac deserve the credit for expanding the mathematical scheme outlined above into a consistent and practically usable theory. These investigators observed in the first place that the quantum conditions can be written as commutation relations between the matrices representing the momenta and the coordinates of the electrons, to yield the equations (p_r, momentum matrices; q_r, coordinate matrices):

$$p_r q_s - q_s p_r = \frac{h}{2\pi i}\delta_{rs}, \quad q_r q_s - q_s q_r = 0, \quad p_r p_s - p_s p_q = 0$$

$$\delta_{rs} = \begin{cases} 1 \text{ for } r = s \\ 0 \text{ for } r \neq s \end{cases}$$

By means of these commutation relations they were able to detect in quantum mechanics as well the laws which were fundamental to classical mechanics: the invariability in time of energy, momentum, and angular momentum.

The mathematical scheme so derived thus ultimately bears an extensive formal similarity to that of the classical theory, from which it differs outwardly by the commutation relations which, moreover, enabled the equations of motion to be derived from the Hamiltonian function.

In the physical consequences, however, there are very profound differences between quantum mechanics and classical mechanics which impose the need for a thorough discussion of the physical interpretation of quantum mechanics. As hitherto defined, quantum mechanics enables the radiation emitted by the atom, the energy values of the stationary states, and other parameters characteristic for the stationary states to be treated. The theory hence complies with the experimental data contained in atomic spectra. In all those cases, however, where a visual description is required of a transient event, e.g. when interpreting Wilson photographs, the formalism of the theory does not seem to allow an adequate representation of the experimental state of affairs. At this point Schrödinger's wave mechanics, meanwhile developed on the basis of de Broglie's theses, came to the assistance of quantum mechanics.

In the course of the studies which Mr. Schrödinger will report here himself he converted the determination of the energy values of an atom into an eigenvalue problem defined by a boundary-value problem in the coordinate space of the particular atomic system. After Schrödinger had shown the mathematical equivalence of wave mechanics, which he had discovered, with quantum mechanics, the fruitful combination of these two different areas of physical ideas resulted in an extraordinary broadening and enrichment of the formalism of the quantum theory. Firstly it was only wave mechanics which made possible the mathematical treatment of complex atomic systems, secondly analysis of the connection between the two theories led to what is known as the transformation theory developed by Dirac and Jordan. As it is impossible within the limits

110

of the present lecture to give a detailed discussion of the mathematical structure of this theory, I should just like to point out its fundamental physical significance. Through the adoption of the physical principles of quantum mechanics into its expanded formalism, the transformation theory made it possible in completely general terms to calculate for atomic systems the probability for the occurrence of a particular, experimentally ascertainable, phenomenon under given experimental conditions. The hypothesis conjectured in the studies on the radiation theory and enunciated in precise terms in Born's collision theory, namely that the wave function governs the probability for the presence of a corpuscle, appeared to be a special case of a more general pattern of laws and to be a natural consequence of the fundamental assumptions of quantum mechanics. Schrödinger, and in later studies Jordan, Klein, and Wigner as well, had succeeded in developing as far as permitted by the principles of the quantum theory de Broglie's original concept of visualizable matter waves occurring in space and time, a concept formulated even before the development of quantum mechanics. But for that the connection between Schrödinger's concepts and de Broglie's original thesis would certainly have seemed a looser one by this statistical interpretation of wave mechanics and by the greater emphasis on the fact that Schrödinger's theory is concerned with waves in multidimensional space. Before proceeding to discuss the explicit significance of quantum mechanics it is perhaps right for me to deal briefly with this question as to the existence of matter waves in three-dimensional space, since the solution to this problem was only achieved by combining wave and quantum mechanics.

A long time before quantum mechanics was developed Pauli had inferred from the laws in the Periodic System of the elements the well-known principle that a particular quantum state can at all times be occupied by only a single electron. It proved possible to transfer this principle to quantum mechanics on the basis of what at first sight seemed a surprising result: the entire complex of stationary states which an atomic system is capable of adopting breaks down into definite classes such that an atom in a state belonging to one class can never change into a state belonging to another class under the action of whatever perturbations. As finally clarified beyond question by the studies of Wigner and Hund, such a class of states is characterized by a defmite symmetry characteristic of the Schrödinger eigenfunction with respect to the transposition of the coordinates of two electrons. Owing to the fundamental identity of electrons, any external perturbation of the atom remains unchanged when two electrons are exchanged and hence causes no transitions between states of various classes. The Pauli principle and the Fermi-Dirac statistics derived from it are equivalent with the assumption that only that class of stationary states is achieved in nature in which the eigenfunction changes its sign when two electrons are exchanged. According to Dirac, selecting the symmetrical system of terms would lead not to the Pauli principle, but to Bose-Einstein electron statistics.

Between the classes of stationary states belonging to the Pauli principle or to Bose-Einstein statistics, and de Broglie's concept of matter waves there is a peculiar relation. A spatial wave phenomenon can be treated according to the principles of the quantum theory by analysing it using the Fourier theorem and then applying to the individual Fourier component of the wave motion, as a system having one degree of freedom, the normal laws of quantum mechanics. Applying this procedure for treating wave phenomena by the quantum theory, a procedure that has also proved fruitful in Dirac's studies of the theory of radiation, to de Broglie's matter waves, exactly the same results are obtained as in treating a whole complex of material particles according to quantum mechanics and selecting the symmetrical system of terms. Jordan and Klein hold that the two methods are mathematically equivalent even if allowance is also made for the interaction of the electrons, i.e. if the field energy originating from the continuous space charge is included in the calculation in de Broglie's wave theory. Schrödinger's considerations of the energy-momentum tensor assigned to the matter waves can then also be adopted in this theory as consistent components of the formalism. The studies of Jordan and Wigner show that modifying the commutation relations underlying this quantum theory of waves results in a formalism equivalent to that of quantum mechanics based on the assumption of Pauli's exclusion principle.

These studies have established that the comparison of an atom with a planetary system composed of nucleus and electrons is not the only visual picture of how we can imagine the atom. On the contrary, it is apparently no less correct to compare the atom with a charge cloud and use the correspondence to the formalism of the quantum theory borne by this concept to derive qualitative conclusions about the behaviour of the atom. However, it is the concern of wave mechanics to follow these consequences. Reverting therefore to the formalism of quantum mechanics; its application to physical problems is justified partly by the original basic assumptions of the theory, partly by its expansion in the transformation theory on the basis of wave mechanics, and the question is now to expose the explicit significance of the theory by comparing it with classical physics.

In classical physics the aim of research was to investigate objective processes occurring in space and time, and to discover the laws governing their progress from the initial conditions. In classical physics a problem was considered solved when a particular phenomenon had been proved to occur objectively in space and time, and it had been shown to obey the general rules of classical physics as formulated by differential equations. The manner in which the knowledge of each process had been acquired, what observations may possibly have led to its experimental determination, was completely immaterial, and it was also immaterial for the consequences of the classical theory, which possible observations were to verify the predictions of the theory. In the quantum theory, however, the situation is completely different. The very fact that the formalism of quantum mechanics cannot be interpreted as visual description of a phenomenon occurring in space and time shows that quantum mechanics is in no way concerned with the objective determination of space-time phenomena. On the contrary, the formalism of quantum mechanics should be used in such a way that the probability for the outcome of a further experiment may be concluded from the determination of an experimental situation in an atomic system, providing that the system is subject to no perturbations other than those necessitated by performing the two experiments. The fact that the only definite known result to be ascertained after the fullest possible experimental investigation of the system is the probability for a certain outcome of a second experiment shows, however, that each observation must entail a discontinuous change in the formalism describing the atomic process and therefore also a discontinuous change in the physical phenomenon itself. Whereas in the classical theory the kind of observation has no bearing on the event, in the quantum theory the disturbance associated with each observation of the atomic phenomenon has a decisive role. Since, furthermore, the result of an observation as a rule leads only to assertions about the probability of certain results of subsequent observations, the fundamentally unverifiable part of each perturbation must, as shown by Bohr, be decisive for the non-contradictory operation of quantum mechanics. This difference between classical and atomic physics is understandable, of course, since for heavy bodies such as the planets moving around the sun the pressure of the sunlight which is reflected at their surface and which is necessary for them to be observed is negligible; for the smallest building units of matter, however, owing to their low mass, every observation has a decisive effect on their physical behaviour.

The perturbation of the system to be observed caused by the observation is also an important factor in determining the limits within which a visual description of atomic phenomena is possible. If there were experiments which permitted accurate measurement of all the characteristics of an atomic system necessary to calculate classical motion, and which, for example, supplied accurate values for the location and velocity of each electron in the system at a particular time, the result of these experiments could not be utilized at all in the formalism, but rather it would directly contradict the formalism. Again, therefore, it is clearly that fundamentally unverifiable part of the perturbation of the system caused by the measurement itself which hampers accurate ascertainment of the classical characteristics and thus permits quantum mechanics to be applied. Closer examination of the formalism shows that between the accuracy with which the location of a particle can be ascertained and the accuracy with which its momentum can simultaneously be known, there is a relation according to which the product of the probable errors in the measurement of the location and momentum is invariably at least as large as Planck's constant divided by 4π. In a very general form, therefore, we should have

$$\Delta p \, \Delta q \geqslant \frac{h}{4\pi},$$

where p and q are canonically conjugated variables. These uncertainty relations for the results of the measurement of classical variables form the necessary conditions for enabling the result of a measurement to be expressed in the formalism of the quantum theory. Bohr has shown in a series of examples how the perturbation necessarily associated with each observation indeed ensures that one cannot go below the limit set by the uncertainty relations. He contends that in the final analysis an uncertainty introduced by the concept of measurement itself is responsible for part of that perturbation remaining fundamentally unknown. The experimental determination of whatever space-time events invariably necessitates a fixed frame – say the system of coordinates in which the observer is at rest – to which all measurements are referred. The assumption that this frame is "fixed" implies neglecting its momentum from the outset, since "fixed" implies nothing other, of course, than that any transfer of momentum to it will evoke no perceptible effect. The fundamentally necessary uncertainty at this point is then transmitted via the measuring apparatus into the atomic event.

Since in connection with this situation it is tempting to consider the possibility of eliminating all uncertainties by amalgamating the object, the measuring apparatuses, and the observer into one quantum-mechanical system, it is important to emphasize that the act of measurement is necessarily visualizable, since, of course, physics is ultimately only concerned with the systematic description of space-time processes. The behaviour of the observer as well as his measuring apparatus must therefore be discussed according to the laws of classical physics, as otherwise there is no further physical problem whatsoever. Within the measuring apparatus, as emphasized by Bohr, all events in the sense of the classical theory will therefore be regarded as determined, this also being a necessary condition before one can, from a result of measurements, unequivocally conclude what has happened. In quantum theory, too, the scheme of classical physics which objectifies the results of observation by assuming in space and time processes obeying laws is thus carried through up to the point where the fundamental limits are imposed by the unvisualizable character of the atomic events symbolized by Planck's constant. A visual description for the atomic events is possible only within certain limits of accuracy – but within these limits the laws of classical physics also still apply. Owing to these limits of accuracy as defined by the uncertainty relations, moreover, a visual picture of the atom free from ambiguity has not been determined. On the contrary the corpuscular and the wave concepts are equally serviceable as a basis for visual interpretation.

The laws of quantum mechanics are basically statistical. Although the parameters of an atomic system are determined in their entirety by an experiment, the result of a future observation of the system is not generally accurately predictable. But at any later point of time there are observations which yield accurately predictable results. For the other observations only the probability for a particular outcome of the experiment can be given. The degree of certainty which still attaches to the laws of quantum mechanics is, for example, responsible for the fact that the principles of conservation for energy and momentum still hold as strictly as ever. They can be checked with any desired accuracy and will then be valid according to the accuracy with which they are checked. The statistical character of the laws of quantum mechanics, however, becomes apparent in that an accurate study of the energetic conditions renders it impossible to pursue at the same time a particular event in space and time.

For the clearest analysis of the conceptual principles of quantum mechanics we are indebted to Bohr who, in particular, applied the concept of complementarity to interpret the validity of the quantum-mechanical laws. The uncertainty relations alone afford an instance of how in quantum mechanics the exact knowledge of one variable can exclude the exact knowledge of another. This complementary relationship between different aspects of one and the same physical process is indeed characteristic for the whole structure of quantum mechanics. I had just mentioned that, for example, the determination of energetic relations excludes the detailed description of space-time processes. Similarly, the study of the chemical properties of a molecule is complementary to the study of the motions of the individual electrons in the molecule, or the observation of interference phenomena complementary to the observation of individual light quanta. Finally, the areas of validity of classical and quantum mechanics can be marked off one from the other as follows: Classical physics represents that striving to learn about Nature in which essentially we seek to draw conclusions about objective processes from observations and so ignore the consideration of the influences which every observation has on the object to be observed; classical physics, therefore, has its limits at the point from which the influence of the observation on the event can no longer be ignored. Conversely, quantum mechanics makes possible the treatment of atomic processes by partially foregoing their space-time description and objectification.

So as not to dwell on assertions in excessively abstract terms about the interpretation of quantum mechanics, I would like briefly to explain with a well-known example how far it is possible through the atomic theory to achieve an understanding of the visual processes with which we are concerned in daily life. The interest of research workers has frequently been focused on the phenomenon of regularly shaped crystals suddenly forming from a liquid, e.g. a supersaturated salt solution. According to the atomic theory the forming force in this process is to a certain extent the symmetry characteristic of the solution to Schrödinger's wave equation, and to that extent crystallization is explained by the atomic theory. Nevertheless this process retains a statistical and – one might almost say – historical element which cannot be further reduced: even when the state of the liquid is completely known before crystallization, the shape of the crystal is not determined by the laws of quantum mechanics. The formation of regular shapes is just far more probable than that of a shapeless lump. But the ultimate shape owes its genesis partly to an element of chance which in principle cannot be analysed further.

Before closing this report on quantum mechanics, I may perhaps be allowed to discuss very briefly the hopes that may be attached to the further development of this branch of research. It would be superfluous to mention that the development must be continued, based equally on the studies of de Broglie, Schrödinger, Born, Jordan, and Dirac. Here the attention of the research workers is primarily directed to the problem of reconciling the claims of the special relativity theory with those of the quantum theory. The extraordinary advances made in this field by Dirac about which Mr. Dirac will speak here, meanwhile leave open the question whether it will be possible to satisfy the claims of the two theories without at the same time determining the Sommerfeld fine-structure constant. The attempts made hitherto to achieve a relativistic formulation of the quantum theory are all based on visual concepts so close to those of classical physics that it seems impossible to determine the fine-structure constant within this system of concepts. The expansion of the conceptual system under discussion here should, furthermore, be closely associated with the further development of the quantum theory of wave fields, and it appears to me as if this formalism, notwithstanding its thorough study by a number of workers (Dirac, Pauli, Jordan, Klein, Wigner, Fermi) has still not been completely exhausted. Important pointers for the further development of quantum mechanics also emerge from the experiments involving the structure of the atomic nuclei. From their analysis by means of the Gamow theory, it would appear that between the elementary particles of the atomic nucleus forces are at work which differ somewhat in type from the forces determining the structure of the atomic shell; Stem's experiments seem, furthermore, to indicate that the behaviour of the heavy elementary particles cannot be represented by the formalism of Dirac's theory of the electron. Future research will thus have to be pre- pared for surprises which may otherwise come both from the field of experience of nuclear physics as well as from that of cosmic radiation. But however the development proceeds in detail, the path so far traced by the quantum theory indicates that an understanding of those still unclarified features of atomic physics can only be acquired by foregoing visualization and objectification to an extent greater than that customary hitherto. We have probably no reason to regret this, because the thought of the great epistemological difficulties with which the visual atom concept of earlier physics had to contend gives us the hope that the abstracter atomic physics developing at present will one day fit more harmoniously into the great edifice of Science.

THE FUNDAMENTAL IDEA OF
WAVE MECHANICS

NOBEL LECTURE DELIVERED AT STOCKHOLM ON
DECEMBER 12TH, 1933

ERWIN SCHRÖDINGER

When a ray of light passes through an optical instrument,
such as a telescope or a photographic lens, it undergoes a
change of direction as it strikes each refractive or reflective
surface. We can describe the path of the light ray once
we know the two simple laws which govern the change

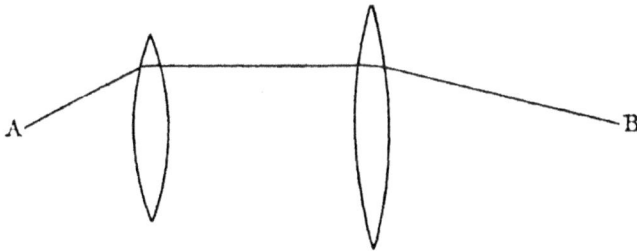

Fig. 1

of direction. One of these is the law of refraction which
was discovered by Snell about three hundred years ago;
and the other is the law of reflection, which was known
to Archimedes nearly two thousand years before. Figure
1 gives a simple example of a ray, $A - B$, passing through
two lenses and undergoing a change of direction at each of
the four surfaces in accordance with Snell's law.

From a much more general point of view, Fermat summed up the whole career of a light ray. In passing through media of varying optical densities light is propagated at correspondingly varying speeds, and the path which it follows is such as would have to be chosen by the light if it had the purpose of arriving *within the quickest possible time* at the destination which it actually reaches. (Here it may be remarked, in parenthesis, that any two points

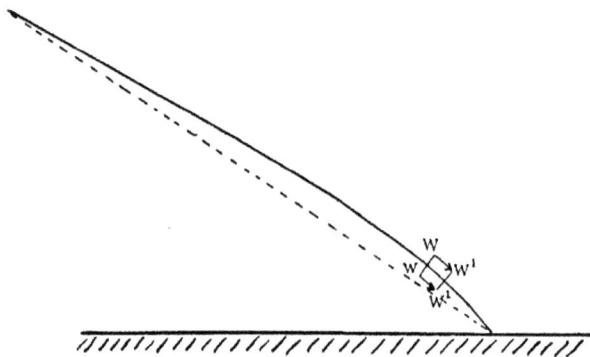

Fig. 2

along the path of the light ray can be chosen as the points of departure and arrival respectively.) Any deviation from the path which the ray has actually chosen would mean a delay. This is Fermat's famous *Principle of Minimum Light Time*. In one admirably concise statement it defines the whole career of a ray of light, including also the more general case where the nature of the medium does not change suddenly but alters gradually from point to point. The atmosphere surrounding our earth is an example of this. When a ray of light, coming from outside, enters the earth's atmosphere the ray travels more slowly as it penetrates into deeper and increasingly denser layers. And although the difference in the speed of propagation is extremely small, yet under these circumstances Fermat's Principle demands that the ray of light must bend earthwards (see Fig. 2), because by doing so it travels for a somewhat longer time in the higher "speedier" layers and comes sooner to its destination than if it were to choose the straight and shorter way (the dotted line in Fig. 2, the small quadrangle WWW^1W^1 to be ignored for the present). Most people will have noticed how the sun no longer presents the shape of a circular disc when it is low on the horizon, but is somewhat flattened, its vertical diameter appearing shortened. That phenomenon is caused by the bending of the light rays as they traverse the earth's atmosphere.

According to the wave theory of light, what we call light rays have, correctly speaking, only a fictitious meaning. They are not the physical tracks of any particles of light, but a purely mathematical construction. The mathematician calls them "orthogonal trajectories" of the wavefronts, that is lines which at every point run at right angles to the wave-surface. Hence they point in the direction in which the light is propagated and, as it were, guide the light's propagation. (See Fig. 3, which represents the sim-

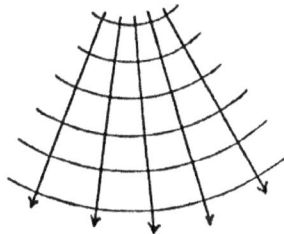

Fig. 3

plest case of concentric spherical wavefronts and the corresponding rectilinear rays, while Fig. 4 illustrates the case of bent rays.) It seems strange that a general principle of such great importance as that of Fermat should be stated directly in reference to these mathematical lines, which are only a mental construction, and not in reference to the wave-fronts themselves. One might therefore be inclined to take it merely for a mathematical curiosity. But that would be a serious mistake. For only from the viewpoint

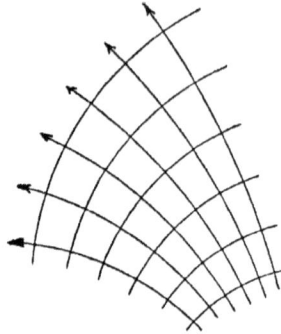

Fig. 4

of the wave theory does this principle become directly and immediately intelligible and cease to be a miracle. What we called *bending* of the light ray presents itself to the wave theory a sa *turning* of the wave-front, and is much more readily understood. For that is just what we must expect in consequence of the fact that neighbouring portions of the wave-fronts advance at various speeds; the turning is effected in the same way as with a company of soldiers marching in line, who are ordered to "right wheel". Here the soldiers in each rank take steps of varying lengths, the man on the right wing taking the shortest steps and the man on the left taking the longest. In the case of atmospheric refraction (Fig. 2) consider a small portion WW of the wave-surface. This portion must necessarily perform a "right wheel" towards W^1W^1, because its left part is in the somewhat higher and rarer air and therefore is moving forward faster than the right, which is in the deeper layer.[1] Now in examining the case more closely it is found that the statement made in Fermat's Principle is virtually

---

[1] In passing, I may call attention to a point in which Snell's concept fails. A ray of light emitted horizontally ought to remain horizontal, because in the horizontal direction the index of refraction does not vary. But, as a matter of fact, a horizontal ray is deflected to a greater degree than any other. According to the concept of the "wheeling" wave-front, this is obvious.

identical with the trivial and obvious assertion that, because the velocity of light varies from point to point, the wave-front must turn, as in the instance I have referred to. I cannot prove that here; but I shall try to show that it is quite reasonable.

Let us revert to the row of soldiers marching in line. To prevent the front rank losing its perfect alignment, let us suppose that a long pole is placed abreast of the men and that each man holds it firmly with his hand against his chest. No word of command as to direction is given, but simply the order that each man must march or run as fast as he can. If the condition of the ground slowly changes from place to place, then either the left or the right section of the line advances more quickly than the other, and this inevitably produces quite spontaneously a wheeling of the whole line to the right or left respectively. After a time it will be noticed that the line of advance, when looked upon as a whole, is not straight, but shows a definite curvature. Now this curved route is precisely the one along which the soldiers reach any place on their way *in the shortest possible time,* taking into account the nature of the ground. Although this may seem remarkable, there is actually nothing strange about it for, after all, by hypothesis, each soldier has done his best to travel as quickly as possible. And it may be further noticed that the bending will always have taken place in the direction towards which the condition of the ground underfoot is less favourable; so that finally it will appear as if the marchers had purposely avoided unfavourable conditions by making a detour around those regions where they would have found their forward pace slackened.

Thus Fermat's Principle directly appears as the *trivial quintessence* of the wave theory. Hence it was a very remarkable event when Hamilton one day made the theoretical discovery that the orbit of a mass point moving in a field of force (for instance, of a stone thrown in the gravitational field of the earth or of some planet in its course around the sun) is governed by a very similar general principle, which thenceforth bore the name of the discoverer and made him famous. Although Hamilton's principle does not precisely consist in the statement that the mass point chooses the quickest way, yet it states something *so* similar – that is to say, it is *so closely* analogous to the principle of minimum light time – that one is faced with a puzzle. It seemed as if Nature had effected exactly the same thing twice, but in two very different ways – once, in the case of light, through a fairly transparent wave-mechanism, and on the other occasion, in the case of mass points by methods which were utterly mysterious, unless one was prepared to believe in some underlying undulatory character in the second case also. But at first sight this idea seemed impossible. For the laws of mechanics had at that time only been established and confirmed experimentally on a large scale for bodies of visible and (in the case of the planets) even huge dimensions which played the role of "mass points", so that something like an "undulatory nature" here appeared to be inconceivable.

The smallest and ultimate constructive elements in the constitution of matter, which we now call "mass points" in a much more particular sense, were at that time purely hypothetical. It was not until the discovery of radioactivity that the process of steadily refining our methods of measurement inaugurated a more detailed investigation of these corpuscles or particles; the development was crowned by C. T. R. Wilson's highly ingenious method, which succeeded in taking snapshots of the track of a single particle and measuring it very accurately by means of stereometric photographs. As far as the measurements go they confirm, in the case of corpuscles, the validity of the same mechanical laws that hold on a large scale, as with planets, etc. Moreover, it was found that neither the molecules nor the atoms are to be considered as the ultimate building stones

of matter, but that the atom itself is an extremely complicated composite system. Definite ideas were formed of the
way in which atoms are composed of corpuscles, leading
to models that closely resembled the celestial planetary
system. And it was natural that in the theoretical construction of these tiny systems the attempt was at first
made to use the same laws of motion as had been so successfully proved to hold good on a large scale. In other
words we endeavoured to conceive the "inner" life of the
atom in terms of Hamiltonian mechanics, which, as I have
said, have their culmination in the Hamiltonian principle.
Meanwhile the very close analogy between the latter and
Fermat's optical principle had been almost entirely forgotten. Or if any thought was given to this at all, the analogy
was looked upon as merely a curious feature of the mathematical theory of the subject.

Now it is very difficult, without going closely into details, to give a correct notion of the success or failure en-
countered in the attempt to explain the structure of matter
by this picture of the atom which was based on classical
mechanics. On the one hand the Hamiltonian principle
directly proved itself to be the truest and most reliable
guide; so much so as to be considered absolutely indispensable. On the other hand, in order to account for certain facts, one had to tolerate the "rude intrusion" (*groben
Eingriff*) of quite new and incomprehensible postulates,
which were called quantum conditions and quantum postulates. These were gross dissonances in the symphony of
classical mechanics – and yet they were curiously chiming
in with it, as if they were being played on the same instrument. In mathematical language, the situation may
be stated thus: The Hamiltonian principle demands only
that a certain integral must be a minimum, without laying
down the numerical value of the minimum in this demand;
the new postulates require that the numerical value of the
minimum must be a whole multiple of a universal constant,

which is Planck's Quantum of Action. But this, only in parenthesis. The situation was rather hopeless. If the old mechanics had failed entirely, that would have been tolerable, for thus the ground would have been cleared for a new theory. But as it was, we were faced with the difficult problem of saving its *soul*, whose breath could be palpably detected in this microcosm, and at the same time persuading it, so to speak, not to consider the quantum conditions "rude intruders" but something arising out of the inner nature of the situation itself.

The way out of the difficulty was actually (though unexpectedly) found in the possibility I have already mentioned, namely, that in the Hamiltonian Principle we might also assume the manifestation of a "wave-mechanism", which we supposed to lie at the basis of events in point mechanics, just as we have been long accustomed to acknowledge it in the phenomena of light and in the governing principle enunciated by Fermat. By this, of course, the individual "path" of a mass point absolutely loses its inherent physical significance and becomes something fictitious, just as the individual light ray. Yet the "soul" of the theory, the minimum principle, not only remains inviolate but we could never even reveal its true and simple meaning (as was stated above), *without* introducing the wave theory. The new theory is in reality no *new* theory but is a thorough organic expansion and development, one might almost say, merely a restatement of the old theory in more subtle terms.

But how could this new and more "subtle" interpretation lead to results that are appreciably different? When applied to the atom, how could it solve any difficulty which the old interpretation could not cope with? How can this new standpoint make that "rude intruder" (*groben Eingriff*) not merely tolerable but even a welcome guest and part of the household, as it were?

These questions, too, can best be elucidated by reference to the analogy with optics. Although I have asserted, and with good reason, that Fermat's principle is the quintessence of the wave theory of light, yet that principle is not such as to render superfluous a more detailed study of wave processes. The optical phenomena of *diffraction* and *interference* can be understood only when we follow up the particulars of the wave process; because these phenomena depend not merely upon where the wave finally arrives but also on whether at a given moment it arrives there as a wave-crest or a wave-trough. To the older and cruder methods of investigation interference phenomena appeared as only small details and escaped observation. But as soon as they were observed and properly accounted for by means of the undulatory theory, quite a number of experimental devices could be easily arranged in which the undulatory character of light was prominently displayed, not only in the finer details but also in the general character of the experiment.

To explain this I shall bring forward two examples: the first is that of an optical instrument, such as a telescope or a microscope. With such an instrument we aim at obtaining a sharp image. This means that we endeavour to focus all the rays emitted from an object point and reunite them at what is called the image point (see Fig. 1a).

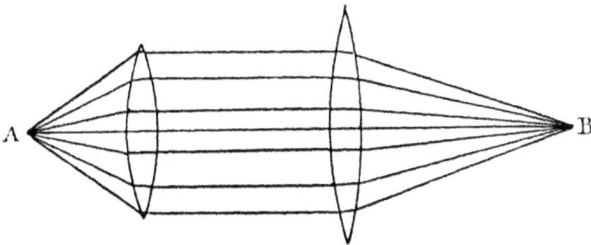

Fig. 1a

Formerly it was thought that the difficulties which stood in the way were only those of geometrical optics, which are actually very considerable. Later it turned out that even in the best constructed instruments lack of precise focussing was considerably greater than might have been expected if in reality each ray, independently of its neighbouring ray, followed Fermat's principle exactly. The light which is emitted from a luminous point and received by an instrument does not focus at an exact point after it has passed the instrument. Instead of this, it covers a small circular area, which is called the diffraction image

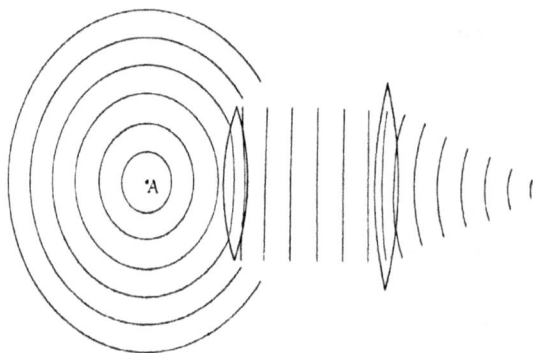

Fig. 1b

and which is mostly circular only because the diaphragms and the circumference of the lenses are usually circular. For diffraction results from the fact that the instrument cannot possibly receive the whole of the spherical waves which are emitted from a luminous point. The borders of the lenses, and sometimes the diaphragms, cut off a part of the wave surface (Fig. 1b) and – if I may use a somewhat crude expression – the torn edges of the wound prevent an exact focus at a point and bring about the indistinctness or blurring of the image. This blurring is closely connected with the *wave-length* of the light and is absolutely unavoidable, owing to this deeply-seated theoretical connection. This phenomenon, originally scarcely noticed, now completely governs and inescapably limits the efficiency of the modern microscope, all the other causes of a lack of distinctness in the image having been successfully overcome. With respect to details, which are not much more coarse-grained than the wave-length of light, the optical image can only reach a distant similarity to the original, and none at all whenever the structural details in the object are *finer* than the wave-length.

The second example is of a simpler nature. Let us take a tiny source of light, just a point only. If we place an opaque body between it and a screen we find a shadow thrown on the screen. To construct the shadow theoretically we should follow each ray of light emitted from the point and should ascertain whether the opaque body prevents it from reaching the screen. The *rim* of the shadow is formed by those light rays which just graze and pass by the outline of the opaque body. But it can be shown by experiment that even where the light source is made as minute as possible, and the outline of the opaque body as sharp as possible, the outer rim of the shadow cast by the opaque body on the screen is not really sharp. The cause of this is again the same as in the former example. The wave-front is split, as it were (Fig. 5), by the outline of the opaque body; and the traces of this lesion blur the rim of the shadow. This would be inexplicable if the individual light rays were independent in themselves and travelled independently with no reference to one another.

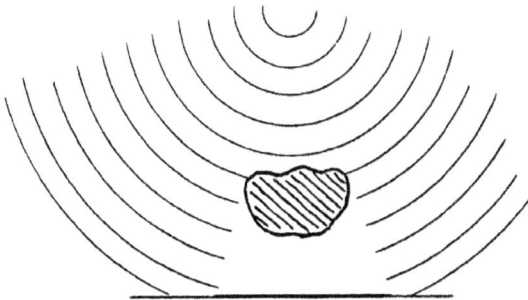

Fig. 5

This phenomenon, which is also called *diffraction,* is

131

generally speaking not very noticeable where larger bodies are concerned. But if the opaque body which throws the shadow be very small, at least in one dimension, then the diffraction has two effects, first, nothing like a true shadow is produced and, secondly – which is far more striking – the tiny body seems to be glowing with its own light and emitting rays in all directions (predominantly, however, at very narrow angles with the incoming rays). Everybody is familiar with the so-called "motes" that appear in the track of a sunbeam entering a dark room. In the same way the filigree of tiny strands and cobwebs that appear around the brow of a hill behind which the sun is hidden, or even the hair of a person standing against the sun, sometimes glows marvellously with diffracted light. The visibility of smoke and fog is due to the same phenomenon. In all these cases the light does not really issue from the opaque body itself but from its immediate surroundings, that is to say, from the area in which the body produces a con-siderable perturbation of the incident wave-fronts. It is interesting, and for what follows very important, to note that the area of perturbation is always and in every di-rection at least as large as one or a few wave-lengths, no matter how small the opaque body may be. Here again, therefore, we see the close relation between wavelength and the phenomenon of diffraction. Perhaps this can be more palpably illustrated by reference to another wave process, namely, that of sound. Here on account of the much longer wave-length, which extends into centimetres and metres, the shadow loses all distinctness and the diffraction pre-dominates to a degree that is of practical importance. We can distinctly *hear* a call from behind a high wall or around the corner of a solid building, although we cannot *see* the person who calls.

Let us now return from optics to mechanics and try to develop the analogy fully. The optical parallel of the *old* mechanics is the method of dealing with isolated rays of light, which are supposed not to influence one another. The new wave mechanics has its parallel in the undulatory theory of light. The advantage of changing from the old concept to the new must obviously consist in clearer insight into diffraction phenomena, or rather into something that is strictly analogous to the diffraction of light, although ordinarily even less significant; for otherwise the old mechanics could not have been accepted as satisfactory for so long a time. But it is not difficult to conjecture the conditions in which the neglected phenomenon must become very prominent, entirely dominate the mechanical process and present problems that are insoluble under the old concept. This occurs inevitably *whenever the entire mechanical system is comparable in its extension with the wave-lengths of "material waves,"* which play the same role in mechanical processes as light waves do in optics.

That is the reason why, in the tiny system of the atom, the old concept is bound to fail. In mechanical phenomena on a large scale it will retain its validity as an excellent approximation, but it must be replaced by the new concept if we wish to deal with the fine interplay which takes place within regions of the order of magnitude of only one or a few wave-lengths. It was amazing to see all the strange additional postulates, which I have mentioned, arising quite automatically from the new undulatory concept, whereas they had to be artificially grafted onto the old one in order to make it fit in with the internal processes of the atom and yield a tolerable explanation of its (the atoms) actually observed manifestations.

In this connection, it is, of course, of outstanding importance that the diameter of the atom and the wave-length of these hypothetical "material" waves should be very nearly of the same order of magnitude. And you will undoubtedly ask whether we are to consider it as purely an accident that in the progressive analysis of the structure of matter we should just here encounter the wave-length order of magnitude, or whether this can be explained. Is there any further evidence of the equality in question? Since the material waves are an entirely new requisite of this theory, which had not been hitherto discerned elsewhere, one might suspect that it is merely a question of suitable *assumption* as to their wave-length, an assumption forced upon us in order to support the preceding arguments.

Well, the coincidence between the two orders of magnitude is by no means a mere accident, and there is no necessity to make any particular assumption in this regard: the coincidence follows naturally from the theory, on account of the following remarkable circumstances. Let us begin by stating that Rutherford's and Chadwick's experiments on the dispersion of Alpha rays have firmly established the fact that the heavy *nucleus* of the atom is very much smaller than the atom, which justifies us in treating it as a point-like centre of attraction in all the argument which follows. Instead of the *electron* we introduce hypothetical waves, the wave-length of which is left an open question as yet, because we do not know anything about it. It is true that this introduces into our calculations a symbol, say a, which represents a number as yet undefined. But in such calculations we are accustomed to that sort of thing and it does not hinder us from inferring that the nucleus of the atom will inevitably produce a sort of diffraction phenomenon of these waves, just like a minute mote does with light waves. Precisely as with light waves, here too the extension of the perturbed area surrounding the nucleus turns out to bear a close relation to the wave-length and to be of the same order of magnitude. Remember that the latter had to be left an open question! But now comes the most important step: *we identify the perturbed area, the diffraction halo, with the atom; the atom being thus regarded as really nothing more than the diffraction phenomenon arising from an electron wave that has been intercepted by the nucleus of the atom.* Thus it is no longer an accident that the size of the atom is of the same order of magnitude as the wave-length. It is in the nature of the case itself. Of course numerically we know neither the one nor the other; because in our calculation there always remains this *one* undefined constant which we have called *a*. It can, however, be determined in two ways, which control one another reciprocally. Either we can choose for *a* that value which will quantitatively account for the observable effects produced by the atom, especially for the emitted spectral lines, which can be measured with extreme accuracy. Or, in the second place, the value of *a* can be adapted in order to give to the diffraction halo the right size, which from other evidence is to be expected for the atom. These two ways of defining *a* (of which the second is, of course, much less definite, because the phrase "size of the atom" is somewhat indefinite) are *in perfect accord with one another.* Thirdly, and finally, it may be remarked that the constant which has remained indeterminate has not really the physical dimension of Length, but of Action, that is, energy multiplied by time. It is, then, very suggestive to assign to it the numerical value of Planck's universal Quantum of Action, which is known with fair accuracy from the laws of heat radiation. The result is that with all desirable exactitude, *we now fall back upon the first (the most exact) method of determining a.* Thus, from the quantitative point of view, the theory answers its purpose with a minimum of new assumptions. It involves a single available constant, to which we only have to assign a numerical value that is already quite familiar to us in the earlier Quantum Theory, in order, first, to give the proper magnitude to the diffraction halos and therewith render possible their identification with the atoms; and, secondly, to calculate with quantitative exactitude all the observable effects produced by the atoms, their radiation of light, the energy required for ionization, etc., etc.

I have tried to explain to you in the simplest possible manner the fundamental concept on which this wave theory of matter is based. Let me confess that, in order to avoid bringing the subject before you in an abstruse form at the very outset, I have embellished it somewhat. Not indeed as regards the thoroughness with which conclusions properly deduced from the theory have been corroborated by experiment, but rather as regards the conceptual simplicity and absence of difficulty in the chain of reasoning which leads to these conclusions. In saying this I do not refer to the mathematical difficulties, which eventually are always trivial, but rather to the conceptual difficulties. Naturally it does not call for a great mental effort to pass from the idea of a path to a system of wave-fronts perpendicular to the path (see Fig. 6). But the wave-surfaces, even when we restrict them to small elements of surface,

Fig. 6

still involve at least a slender *bundle* of possible paths, to all of which they stand in the same relation. According to the traditional idea, in each concrete case one of these paths is singled out as the one "really travelled", in contradistinction to all the other "merely possible" paths. According to the new concept the case is quite different. We are confronted with the profound logical antithesis between

Either this or that (Particle Mechanics)
(aut − aut)

and

This as well as that (Wave Mechanics)
(et − et).

Now this would not be so perplexing if it were really a question of abandoning the old concept and *substituting* the new one for it. But unfortunately that is not the state of affairs. From the standpoint of wave mechanics the innumerable multitude of possible particle paths would be only fictitious and no single one would have the special prerogative of being that actually travelled in the individual case. But, as I have already remarked, we have in some cases actually observed such individual tracks of a particle. The wave theory cannot meet this case, except in a very unsatisfactory way. We find it extraordinarily difficult to regard the track whose trace we actually *see*, only as a slender bundle of equally possible (*gleichberechtigten*) tracks between which the wave-fronts form a lateral connection. And yet these lateral connections are necessary to the understanding of diffraction and interference phenomena, which the very same particles produce before our eyes with equal obviousness – that is to say produce experimentally on a large scale and not only in those concepts of the interior of the atom discussed previously. It is true that we can deal with every concrete individual case without the two contrasted aspects leading to different expectations as to the result of any given experiment. But with the old and cherished and apparently indispensable concepts, such as "really" and "merely possible", we cannot advance. We can never say what really *is* or what really *happens,* but only what is *observable,* in each concrete case. Shall we content ourselves with this as a permanent feature? In principle, yes. It is by no means a new demand to claim that, in principle, the ultimate aim of exact science must be restricted to the description of what is really observable. The question is only whether we must henceforth forego connecting the description, as we did hitherto, with a definite hypothesis as to the real structure of the Universe. Today there is a widespread tendency to insist on this renunciation. But I think that this is taking the matter somewhat too lightly.

I would describe the present state of our knowledge as follows: The light ray, or track of the particle, corresponds to a *longitudinal* continuity of the propagating process (that is to say, *in the* direction of the spreading); the wave-front, on the other hand, to a *transversal* one, that is to say, perpendicular to the direction of spreading. *Both* continuities are undoubtedly real. The one has been proved by photographing the particle tracks, and the other by interference experiments. As yet we have not been able to bring the two together into a uniform scheme. It is only in extreme cases that the transversal – the spherical – continuity or the longitudinal – the ray-continuity shows itself so predominantly that we *believe* we can avail ourselves either of the wave scheme or of the particle scheme alone.

THE STATISTICAL INTERPRETATION OF QUANTUM MECHANICS

NOBEL LECTURE, DECEMBER 11, 1954

MAX BORN

The work, for which I have had the honour to be awarded the Nobel Prize for 1954, contains no discovery of a fresh natural phenomenon, but rather the basis for a new mode of thought in regard to natural phenomena. This way of thinking has permeated both experimental and theoretical physics to such a degree that it hardly seems possible to say anything more about it that has not been already so often said. However, there are some particular aspects which I should like to discuss on what is, for me, such a festive occasion. The first point is this: the work at the Göttingen school, which I directed at that time (1926-1927), contributed to the solution of an intellectual crisis into which our science had fallen as a result of Planck's discovery of the quantum of action in 1900. Today, physics finds itself in a similar crisis – I do not mean here its entanglement in politics and economics as a result of the mastery of a new and frightful force of Nature, but I am considering more the logical and epistemological problems posed by nuclear physics. Perhaps it is well at such a time to recall what took place earlier in a similar situation, especially as these events are not without a definite dramatic flavour.

The second point I wish to make is that when I say that the physicists had accepted the concepts and mode of thought developed by us at the time, I am not quite correct. There are some very noteworthy exceptions, particularly among the very workers who have contributed most to building up the quantum theory. Planck, himself, belonged to the sceptics until he died. Einstein, De Broglie, and Schrödinge have unceasingly stressed the unsatisfactory features of quantum mechanics and called for a return to the concepts of classical, Newtonian physics while proposing ways in which this could be done without contradicting experimental facts. Such weighty views cannot be ignored. Niels Bohr has gone to a great deal of trouble to refute the objections. I, too, have ruminated upon them and believe I can make some contribution to the clarification of the position. The matter concerns the borderland between physics and philosophy, and so my physics lecture will partake of both history and philosophy, for which I must crave your indulgence.

First of all, I will explain how quantum mechanics and its statistical interpretation arose. At the beginning of the twenties, every physicist, I think, was convinced that Planck's quantum hypothesis was correct. According to this theory energy appears in finite quanta of magnitude $h\nu$ in oscillatory processes having a specific frequency ν (e.g. in light waves). Countless experiments could be explained in this way and always gave the same value of Planck's constant h. Again, Einstein's assertion that light quanta have *momentum* $h\nu/c$ (where c is the speed of light) was well supported by experiment (e.g. through the Compton effect). This implied a revival of the corpuscular theory of light for a certain complex of phenomena. The wave theory still held good for other processes. Physicists grew accustomed to this *duality* and learned how to cope with it to a certain extent.

In 1913 Niels Bohr had solved the riddle of *line spec-*

tra by means of the quantum theory and had thereby explained broadly the amazing stability of the atoms, the structure of their electronic shells, and the Periodic System of the elements. For what was to come later, the most important assumption of his teaching was this: an atomic system cannot exist in all mechanically possible states, forming a continuum, but in a series of discrete "stationary" states. In a transition from one to another, the difference in energy $E_m - E_n$ is emitted or absorbed as a light quantum $h\nu_{mn}$ (according to whether E_m is greater or less than E_n). This is an interpretation in terms of energy of the fundamental law of spectroscopy discovered some years before by W. Ritz. The situation can be taken in at a glance by writing the energy levels of the stationary states twice over, horizontally and vertically. This produces a square array

$$
\begin{array}{cccc}
 & E_1 & E_2 & E_3 & \dots \\
E_1 & 11 & 12 & 13 & - \\
E_2 & 21 & 22 & 23 & - \\
E_3 & 31 & 32 & 33 & - \\
 & - & - & - & - \\
\end{array}
$$

in which positions on a diagonal correspond to states, and non-diagonal positions correspond to transitions.

It was completely clear to Bohr that the law thus formulated is in conflict with mechanics, and that therefore the use of the energy concept in this connection is problematical. He based this daring fusion of old and new on his *principle of correspondence*. This consists in the obvious requirement that ordinary classical mechanics must hold to a high degree of approximation in the limiting case where the numbers of the stationary states, the so-called quantum numbers, are very large (that is to say, far to the right and to the lower part in the above array) and the energy changes relatively little from place to place, in fact practically continuously.

Theoretical physics maintained itself on this concept for the next ten years. The problem was this: an harmonic oscillation not only has a frequency, but also an intensity. For each transition in the array there must be a corresponding intensity. The question is how to find this through the considerations of correspondence? It meant guessing the unknown from the available information on a known limiting case. Considerable success was attained by Bohr himself, by Kramers, Sommerfeld, Epstein, and many others. But the decisive step was again taken by Einstein who, by a fresh derivation of Planck's radiation formula, made it transparently clear that the classical concept of intensity of radiation must be replaced by the statistical concept of transition probability. To each place in our pattern or array there belongs (together with the frequency $\nu_{mn} = (E_n - E_m)/h$) a definite prob-ability for the transition coupled with emission or absorption. In Göttingen we also took part in efforts to distil the unknown mechanics of the atom from the experimental results. The logical difficulty became ever sharper. Investigations into the scattering and dispersion of light showed that Einstein's conception of transition probability as a measure of the strength of an oscillation did not meet the case, and the idea of an amplitude of oscillation associated with each transition was indispensable. In this connection, work by Ladenburg,[1] Kramer,[2] Heisenberg,[3] Jordan and me[4] should be mentioned. The art of guessing correct formulae, which deviate from the classical formulae, yet contain them as a limiting case according to the correspondence principle, was brought to a high degree of perfection. A paper of mine, which introduced, for the first time I think, the expression *quantum mechanics* in its title, contains a rather involved formula (still valid today) for the reciprocal disturbance of atomic systems.

Heisenberg, who at that time was my assistant, brought this period to a sudden end.[5] He cut the Gordian knot by means of a philosophical principle and replaced guess-work by a mathematical rule. The principle states that concepts and representations that do not correspond to physically observable facts are not to be used in theoretical description. Einstein used the same principle when, in setting up his theory of relativity, he eliminated the concepts of absolute velocity of a body and of absolute simultaneity of two events at different places. Heisenberg banished the picture of electron orbits with definite radii and periods of rotation because these quantities are not observable, and insisted that the theory be built up by means of the square arrays mentioned above. Instead of describing the motion by giving a coordinate as a function of time, $x(t)$, an array of transition amplitudes x_{mn} should be determined. To me the decisive part of his work is the demand to determine a rule by which from a given

$$\text{array} \begin{bmatrix} x_{11} & x_{12} & \cdots \\ x_{21} & x_{22} & \cdots \\ - & - & \cdots \end{bmatrix}$$

$$\text{the array for the square} \begin{bmatrix} (x^2)_{11} & (x^2)_{12} & \cdots \\ (x^2)_{21} & (x^2)_{22} & \cdots \\ - & - & \cdots \end{bmatrix}$$

can be found (or, more general, the *multiplication rule* for such arrays). By observation of known examples solved by guess-work he found this rule and applied it successfully to simple examples such as the harmonic and anharmonic oscillator.

This was in the summer of 1925. Heisenberg, plagued by hay fever took leave for a course of treatment by the sea and gave me his paper for publication if I thought I could do something with it.

The significance of the idea was at once clear to me and I sent the manuscript to the *Zeitschrift für Physik*. I could not take my mind off Heisenberg's multiplication rule, and after a week of intensive thought and trial I suddenly remembered an algebraic theory which I had learned from my teacher, Professor Rosanes, in Breslau. Such square arrays are well known to mathematicians and, in conjunction with a specific rule for multiplication, are called matrices. I applied this rule to Heisenberg's quantum condition and found that this agreed in the diagonal terms. It was easy to guess what the remaining quantities must be, namely, zero; and at once there stood before me the peculiar formula

$$pq - qp = h/2\pi i.$$

This meant that coordinates q and momenta p cannot be represented by figure values but by symbols, the product of which depends upon the order of multiplication – they are said to be "non-commuting."

I was as excited by this result as a sailor would be who, after a long voyage, sees from afar, the longed-for land, and I felt regret that Heisenberg was not there. I was convinced from the start that we had stumbled on the right path. Even so, a great part was only guess-work, in particular, the disappearance of the non-diagonal elements in the above-mentioned expression. For help in this problem I obtained the assistance and collaboration of my pupil Pascual Jordan, and in a few days we were able to demonstrate that I had guessed correctly. The joint paper by Jordan and myself[6] contains the most important principles of quantum mechanics including its extension to electrodynamics. There followed a hectic period of collaboration among the three of us, complicated by Heisenberg's absence. There was a lively exchange of letters; my contribution to these, unfortunately, have been lost in the political disorders. The result was a three-author paper[7] which brought the formal side of the investigation to a definite conclusion. Before this paper appeared, came the first *dramatic surprise*: Paul Dirac's paper on the same subject.[8] The inspiration afforded by a lecture of Heisenberg's in Cambridge had led him to similar results as we had obtained in Göttingen except that he did not resort to the known matrix theory of the mathematicians, but discovered the tool for himself and worked out the theory of such non-commutating symbols.

The first non-trivial and physically important application of quantum mechanics was made shortly afterwards by W. Pauli[9] who calculated the stationary energy values of the *hydrogen atom* by means of the matrix method and found complete agreement with Bohr's formulae. From this moment onwards there could no longer be any doubt about the correctness of the theory.

What this formalism really signified was, however, by no means clear. Mathematics, as often happens, was cleverer than interpretative thought. While we were still discussing this point there came the *second dramatic surprise*, the appearance of Schrödinger's famous papers.[10] He took up quite a different line of thought which had originated from Louis de Broglie.[11]

A few years previously, the latter had made the bold assertion, supported by brilliant theoretical considerations, that wave-corpuscle duality, familiar to physicists in the case of light, must also be valid for electrons. To each electron moving free of force belongs a plane wave of a definite wavelength which is determined by Planck's constant and the mass. This exciting dissertation by De Broglie was well known to us in Göttingen. One day in 1925 I received a letter from C. J. Davisson giving some peculiar results on the reflection of electrons from metallic surfaces. I, and my colleague on the experimental side, James Franck, at once suspected that these curves of Davisson's were crystal-lattice spectra of De Broglie's electron waves, and we made one of our pupils, Elsasser,[12] to investigate the matter. His result provided the first preliminary confirmation of the idea of De Broglie's, and this was later proved independently by Davisson and Germer[13] and G. P. Thomson[14] by systematic experiments.

But this acquaintance with De Broglie's way of thinking did not lead us to an attempt to apply it to the electronic structure in atoms. This was left to Schrödinger. He extended De Broglie's wave equation which referred to force-free motion, to the case where the effect of force is taken into account, and gave an exact formulation of the *subsidiary conditions*, already suggested by De Broglie, to which the wave function ψ must be subjected, namely that it should be single-valued and finite in space and time. And he was successful in deriving the stationary states of the hydrogen atom in the form of those monochromatic solutions of his wave equation which do not extend to infinity. For a brief period at the beginning of 1926, it looked as though there were, suddenly, two self-contained but quite distinct systems of explanation extant: matrix mechanics and wave mechanics. But Schrödinger himself soon demonstrated their complete equivalence.

Wave mechanics enjoyed a very great deal more popularity than the Göttingen or Cambridge version of quantum mechanics. It operates with a wave function ψ, which in the case of *one* particle at least, can be pictured in space, and it uses the mathematical methods of partial differential equations which are in current use by physicists. Schrödinger thought that his wave theory made it possible to return to deterministic classical physics. He proposed[15] (and he has recently emphasized his proposal anew's), to dispense with the particle representation entirely, and instead of speaking of electrons as particles, to consider them as a continuous density distribution $|\psi|^2$ (or electric density $e|\psi|^2$).

To us in Göttingen this interpretation seemed unacceptable in face of well established experimental facts. At that time it was already possible to count particles by means of scintillations or with a Geiger counter, and to photograph their tracks with the aid of a Wilson cloud chamber.

It appeared to me that it was not possible to obtain a clear interpretation of the ψ-function, by considering bound electrons. I had therefore, as early as the end of 1925, made an attempt to extend the matrix method, which obviously only covered oscillatory processes, in such a way as to be applicable to aperiodic processes. I was at that time a guest of the Massachusetts Institute of Technology in the USA, and I found there in Norbert Wiener an excellent collaborator. In our joint paper[16] we replaced the matrix by the general concept of an operator, and thus made it possible to describe aperiodic processes. Nevertheless we missed the correct approach. This was left to Schrödinger, and I immediately took up his method since it held promise of leading to an interpretation of the ψ-function. Again an idea of Einstein's gave me the lead. He had tried to make the duality of particles – light quanta or photons – and waves comprehensible by interpreting the square of the optical wave amplitudes as probability density for the occurrence of photons. This concept could at once be carried over to the ψ-function: $|\psi|^2$ ought to represent the probability density for electrons (or other particles). It was easy to assert this, but how could it be proved?

The atomic collision processes suggested themselves at this point. A swarm of electrons coming from infinity, represented by an incident wave of known intensity (i.e., $|\psi|^2$), impinges upon an obstacle, say a heavy atom. In the same way that a water wave produced by a steamer causes secondary circular waves in striking a pile, the incident electron wave is partially transformed into a secondary spherical wave whose amplitude of oscillation ψ differs for different directions. The square of the amplitude of this wave at a great distance from the scattering centre determines the relative probability of scattering as

147

a function of direction. Moreover, if the scattering atom itself is capable of existing in different stationary states, then Schrödinger's wave equation gives automatically the probability of excitation of these states, the electron being scattered with loss of energy, that is to say, inelastically, as it is called. In this way it was possible to get a theoretical basis[17] for the assumptions of Bohr's theory which had been experimentally confirmed by Franck and Hertz. Soon Wentzel[18] succeeded in deriving Rutherford's famous formula for the scattering of α-particles from my theory.

However, a paper by Heisenberg,[19] containing his celebrated uncertainty relationship, contributed more than the above-mentioned successes to the swift acceptance of the statistical interpretation of the ψ-function. It was through this paper that the revolutionary character of the new conception became clear. It showed that not only the determinism of classical physics must be abandoned, but also the naive concept of reality which looked upon the particles of atomic physics as if they were very small grains of sand. At every instant a grain of sand has a definite position and velocity. This is not the case with an electron. If its position is determined with increasing accuracy, the possibility of ascertaining the velocity becomes less and vice versa. I shall return shortly to these problems in a more general connection, but would first like to say a few words about the theory of collisions. The mathematical approximation methods which I used were quite primitive and soon improved upon. From the literature, which has grown to a point where I cannot cope with, I would like to mention only a few of the first authors to whom the theory owes great progress: Faxén in Sweden, Holtsmark in Norway[20], Bethe in Germany[21], Mott and Massey in England[22].

Today, collision theory is a special science with its own big, solid textbooks which have grown completely over my head. Of course in the last resort all the modern branches of physics, quantum electrodynamics, the theory of mesons, nuclei, cosmic rays, elementary particles and their trans- formations, all come within range of these ideas and no bounds could be set to a discussion on them.

I should also like to mention that in 1926 and 1927 I tried another way of supporting the statistical concept of quantum mechanics, partly in collaboration with the Russian physicist Fock.[23] In the above-mentioned three-author paper there is a chapter which anticipates the Schrödinger function, except that it is not thought of as a function $y(x)$ in space, but as a function ψ_n of the discrete index $n = 1, 2, \ldots$ which enumerates the stationary states. If the system under consideration is subject to a force which is variable with time, ψ_n becomes also time-dependent, and $|\psi_n|^2$ signifies the probability for the existence of the state n at time t. Starting from an initial distribution where there is only one state, transition probabilities are obtained, and their properties can be examined. What interested me in particular at the time, was what occurs in the adiabatic limiting case, that is, for very slowly changing action. It was possible to show that, as could have been expected, the probability of transitions becomes ever smaller. The theory of transition probabilities was developed independently by Dirac with great success. It can be said that the whole of atomic and nuclear physics works with this system of concepts, particularly in the very elegant form given to them by Dirac.[24] Almost all experiments lead to statements about relative frequencies of events, even when they occur concealed under such names as effective cross section or the like.

How does it come about then, that great scientists such as Einstein, Schrödinger, and De Broglie are nevertheless dissatisfied with the situation? Of course, all these objections are levelled not against the correctness of the formulae, but against their interpretation. Two closely knitted points of view are to be distinguished: the question of determinism and the question of reality.

Newtonian mechanics is deterministic in the following sense:

If the initial state (positions and velocities of all particles) of a system is accurately given, then the state at any other time (earlier or later) can be calculated from the laws of mechanics. All the other branches of classical physics have been built up according to this model. Mechanical determinism gradually became a kind of article of faith: the world as a machine, an automaton. As far as I can see, this idea has no forerunners in ancient and medieval philosophy. The idea is a product of the immense success of Newtonian mechanics, particularly in astronomy. In the 19th century it became a basic philosophical principle for the whole of exact science. I asked myself whether this was really justified. Can absolute predictions really be made for all time on the basis of the classical equations of motion? It can easily be seen, by simple examples, that this is only the case when the possibility of absolutely exact measurement (of position, velocity, or other quantities) is assumed. Let us think of a particle moving without friction on a straight line between two end-points (walls), at which it experiences completely elastic recoil. It moves with constant speed equal to its initial speed v_0 backwards and forwards, and it can be stated exactly where it will be at a given time provided that v_0 is accurately known. But if a small inaccuracy Δv_0, is allowed, then the inaccuracy of prediction of the position at time t is $t\Delta v_0$ which increases with t. If one waits long enough until time $t_c = I/\Delta v_0$ where l is the distance between the elastic walls, the inaccuracy Δx will have become equal to the whole space l. Thus it is impossible to forecast anything about the position at a time which is later than t_c. Thus determinism lapses completely into indeterminism as soon as the slightest inaccuracy in the data on velocity is permitted. Is there any sense – and I mean any physical sense, not metaphys-

150

ical sense – in which one can speak of absolute data? Is one justified in saying that the coordinate $x = \pi$ cm where $\pi = 3.1415\ldots$ is the familiar transcendental number that determines the ratio of the circumference of a circle to its diameter? As a mathematical tool the concept of a real number represented by a nonterminating decimal fraction is exceptionally important and fruitful. As the measure of a physical quantity it is nonsense. If π is taken to the 20th or the 25th place of decimals, two numbers are obtained which are indistinguishable from each other and the true value of π by any measurement. According to the heuristic principle used by Einstein in the theory of relativity, and by Heisenberg in the quantum theory, concepts which correspond to no conceivable observation should be eliminated from physics. This is possible without difficulty in the present case also. It is only necessary to replace statements like $x = \pi$ cm by: the probability of distribution of values of x has a sharp maximum at $x = \pi$ cm; and (if it is desired to be more accurate) to add: of such and such a breadth. In short, ordinary mechanics must also be statistically formulated. I have occupied myself with this problem a little recently, and have realized that it is possible without difficulty. This is not the place to go into the matter more deeply. I should like only to say this: the determinism of classical physics turns out to be an illusion, created by overrating mathematico-logical concepts. It is an idol, not an ideal in scientific research and cannot, therefore, be used as an objection to the essentially indeterministic statistical interpretation of quantum mechanics.

Much more difficult is the objection based on reality. The concept of a particle, e.g. a grain of sand, implicitly contains the idea that it is in a definite position and has definite motion. But according to quantum mechanics it is impossible to determine simultaneously with any desired accuracy both position and velocity (more precisely:

151

momentum, i.e. mass times velocity) . Thus two questions arise: what prevents us, in spite of the theoretical assertion, to measure both quantities to any desired degree of accuracy by refined experiments? Secondly, if it really transpires that this is not feasible, are we still justified in applying to the electron the concept of particle and therefore the ideas associated with it?

Referring to the first question, it is clear that if the theory is correct – and we have ample grounds for believing this – the obstacle to simultaneous measurement of position and motion (and of other such pairs of so-called conjugate quantities) must lie in the laws of quantum mechanics themselves. In fact, this is so. But it is not a simple matter to clarify the situation. Niels Bohr himself has gone to great trouble and ingenuity[25] to develop a theory of measurements to clear the matter up and to meet the most refined and ingenious attacks of Einstein, who repeatedly tried to think out methods of measurement by means of which position and motion could be measured simultaneously and accurately. The following emerges: to measure space coordinates and instants of time, rigid measuring rods and clocks are required. On the other hand, to measure momenta and energies, devices are necessary with movable parts to absorb the impact of the test object and to indicate the size of its momentum. Paying regard to the fact that quantum mechanics is competent for dealing with the interaction of object and apparatus, it is seen that no arrangement is possible that will fulfil both requirements simultaneously. There exist, therefore, mutually exclusive though complementary experiments which only as a whole embrace everything which can be experienced with regard to an object.

This idea of *complementarity* is now regarded by most physicists as the key to the clear understanding of quantum processes. Bohr has generalized the idea to quite different fields of knowledge, e.g. the connection be- tween consciousness and the brain, to the problem of free will, and other basic problems of philosophy. To come now to the last point: can we call something with which the concepts of position and motion cannot be ass ciated in the usual way, a thing, or a particle? And if not, what is the reality which our theory has been invented to describe?

The answer to this is no longer physics, but philosophy, and to deal with it thoroughly would mean going far beyond the bounds of this lecture. I have given my views on it elsewhere.[26] Here I will only say that I am emphatically in favour of the retention of the particle idea. Naturally, it is necessary to redefine what is meant. For this, well-developed concepts are available which appear in mathematics under the name of invariants in transformations. Every object that we perceive appears in innumerable aspects. The concept of the object is the invariant of all these aspects. From this point of view, the present universally used system of concepts in which particles and waves appear simultaneously, can be completely justified.

The latest research on nuclei and elementary particles has led us, however, to limits beyond which this system of concepts itself does not appear to suffice. The lesson to be learned from what I have told of the origin of quantum mechanics is that probable refinements of mathematical methods will not suffice to produce a satisfactory theory, but that somewhere in our doctrine is hidden a concept, unjustified by experience, which we must eliminate to open up the road.

1. R. Ladenburg, *Z. Physik*, 4 (192.1) 451; R. Ladenburg and F. Reiche, *Naturwiss.*, 11 (1923) 584.

2. H. A. Kramers, *Nature*, 113 (1924) 673.

3. H. A. Kramers and W. Heisenberg, *Z. Physik*, 31 (1925) 681.

4. M. Born, *Z. Physik*, 26 (1924) 379; M. Born and P.

Jordan, Z. *Physik*, 33 (1925) 479.

5. W. Heisenberg, Z. *Physik*, 33 (1925) 879.

6. M. Born and P. Jordan, Z. *Physik*, 34 (1925) 858.

7. M. Born, W. Heisenberg, and P. Jordan, Z. *Physik*, 35 (1926) 557.

8. P. A. M. Dirac, *Proc. Roy. Soc.* (London), A 109 (1925) 642.

9. W. Pauli, Z. *Physik*, 36 (1926) 336.

10. E. Schrödinger, *Ann. Physik*, [4] 79 (1926) 361,489,734; 80 (1926) 437; 81(1926) 109.

11. L. de Broglie, Thesis Paris, 1924; *Ann. Phys.* (Paris), [10] 3 (1925) 22.

12. W. Elasser, *Naturwiss.*, 13 (1925) 711.

13. C. J. Davisson and L. H. Germer, *Phys. Rev.*, 30 (1927) 707.

14. G. P. Thomson and A. Reid, *Nature*, 119 (1927) 890; G. P. Thomson, *Proc. Roy. Soc.* (London), A 117 (1928) 600.

15. E. Schrödinger, *Brit. J. Phil. Sci.*, 3 (1952) 109, 233.

16. M. Born and N. Wiener, Z. *Physik*, 36 (1926) 174.

17. M. Born, Z. *Physik*, 37 (1926) 863; 38 (1926) 803; *Göttinger Nachr. Math. Phys. Kl.*, (1926) 146.

18. G. Wentzel, Z. *Physik*, 40 (1926) 590.

19. W. Heisenberg, Z. *Physik*, 43 (1927) 172.

20. H. Faxén and J. Holtsmark, Z. *Physik*, 45 (1927) 307.

21. H. Bethe, *Ann. Physik*, 5 (1930) 325.

22. N. F. Mott, *Proc. Roy. Soc.* (London), A 124 (1929) 422, 425; *Proc. Cambridge Phil. Soc.*, 25 (1929) 304.

23. M. Born, Z. *Physik*, 40 (1926) 167; M. Born and V. Fock, Z. *Physik*, 51 (1928) 165.

24. P. A. M. Dirac, *Proc. Roy. Soc.* (London), A 109 (1925) 642; 110 (1926) 561; 111 (1926) 281; 112 (26) 674.

25. N. Bohr, *Naturwiss.*, 16 (1928) 245; 17 (1929) 483; 21 (1933) 13. "Kausalität und Komplementarität" (Causality and Complementarity), *Die Erkenntnis*, 6 (1936) 293.

26. M. Born, *Phil. Quart.*, 3 (1953) 134; *Physik. Bl.*, 10 (1954) 49.

www.ingramcontent.com/pod-product-compliance
Lightning Source LLC
Chambersburg PA
CBHW071555200326
41519CB00021BB/6760